ANTI-ACCESS WARFARE

ANTI-ACCESS WARFARE

Countering A2/AD Strategies

Sam J. Tangredi

Naval Institute Press
Annapolis, Maryland

Naval Institute Press
291 Wood Road
Annapolis, MD 21402

Library of Congress Cataloging-in-Publication Data
Tangredi, Sam J.
Anti-access warfare : countering A2/AD strategies / Sam J. Tangredi.
pages cm
Summary: "This is the first book to examine the concept of anti-access and area denial warfare, providing a definitive introduction to both conceptual theories and historical examples of this strategy. Also referred to by the acronym "A2/AD," anti-access warfare has been identified in American strategic planning as the most likely strategy to be employed by the People's Republic of China or by the Islamic Republic of Iran in any future conflict with the United States. While previous studies of the subject have emphasized the effects on the joint force and, air forces in particular, this important new study advances the understanding of sea power by identifying the naval roots of the development of the anti-access concept. Rather than arguing against the reliance on maritime forces—presumably because they are no longer survivable—Tangredi maintains that history argues that maritime capabilities are key to countering anti-access operations. "— Provided by publisher.
ISBN 978-1-61251-186-3 (hardback : alk. paper) 1. Sea control. 2. Strategy—Case studies. 3. Unified operations (Military science)—Case studies. 4. Defensive (Military science) I. Title. II. Title: Countering A2/AD strategies.
V165.T38 2013
359.4'22—dc23
2013022627
∞ Print editions meet the requirements of ANSI/NISO z39.48–1992 (Permanence of Paper).
Printed in the United States of America.

21 20 19 18 17 16 9 8 7 6 5 4 3

Contents

Illustrations

Tables

Figure

Preface

This is admittedly a book written with an America-centric point of view. It is a point of view for which I make no apologies because—unlike many of my counterparts in the academic world—I believe that there is no such thing as value-free or value-neutral research in international relations or any other social science. Rather, all social science scholarship—that is the study of humanity—is affected by the personal beliefs, or biases if one wants to use the pejorative term, of the researcher. This is true no matter how inclusive the bibliography or sources of research. Humans can never be neutral about humans. This is not an idea that originated with me; one of my more influential graduate professors, a man who had previously spent his lifetime trying to define foreign policy research as a hard science, taught me this wisdom from his own disappointment. Rather than pretend the research to be value-free, he urged, the researcher should be value-explicit. Making one's biases known allows the reader to place the research and analysis into context and therefore better judge its validity. Accepting this, I can point out my biases very succinctly: I hold a lifelong commission as an officer in service of the United States. I may no longer be serving on active duty, but the national security of the United States and its allies has been my life's work and remains my professional focus. This book is a scholarly work in which all sources available to me and all points of view have been researched, but ultimately the reader must remain aware that my conclusions and recommendations necessarily reflect this lifelong mission.

My naval and academic careers have been full of shipmates, colleagues, teachers, mentors, and friends, many of whom have had direct or indirect effects on my studies of anti-access strategies and related defense issues. It is not mere rhetoric to say that they are too many to name, and I know that acknowledging some means that others will be necessarily unacknowledged. So, instead, I will confine myself to first acknowledging my gratitude to the executive leadership of Strategic Insight—Bob Gray, Jim Francis, and Jim Rubin—who allowed me time off from my regular work responsibilities to complete this book. Most importantly, I give my continuing gratitude to Rev. Dr. Deborah Mariya, veteran of Desert Shield/Desert Storm, and our daughter Mercy, whose access to the future—I pray—will always be peaceful.

Introduction

Anti-access and *area denial* are modern terms referring to war-fighting strategies focused on preventing an opponent from operating military forces near, into, or within a contested region. Today anti-access and area-denial strategies—sometimes combined as *anti-access/area denial* or abbreviated as *A2/AD*—are topics of great debate and are considered primary strategic challenges to the international security objectives of the United States and its allies and partners. However, in addition to anti-access and area denial being modern terms and strategic challenges, I would argue that they constitute an ancient concept—that they are techniques of strategy that have been used throughout military history. They are also historical components of grand strategy. This book is an attempt to illustrate this history, identify the principal elements of the concept, assess their relative success or failure, and—in terms of the policy objectives of the United States and its allies and partners—suggest ways in which future strategic challenges might be overcome.

Denying access to an enemy is a natural objective for any defender and should be considered an integral component of any military campaign. However, the terms anti-access and area denial—as currently used—are specifically meant to denote a strategic approach intended to defend against an opponent that is judged to be of superior strength or skill in overall combat operations. If the opponent is allowed to use this superior strength or skill, it is feared that the defender would likely be defeated at the point of contact. Therefore, the objective of an anti-access or area-denial strategy is to prevent the attacker from bringing its operationally superior force into the contested region or to prevent the attacker from freely operating within the region and maximizing its combat power.

In the terms of one school of military theory, anti-access and area denial can be described as strategies intended to prevent an attacker from being able to bring forces to bear in a strike at a defender's center of gravity. From this perspective, without striking the center of gravity the attacker can never achieve victory. For the defender, the desired result is not just stalemate, but also attrition of the attacker's forces such that the attacker loses over time any ability to make any decisive strike at the center. Using chess as the metaphor, we can see it as a strategy focused on preventing the loss of one's own king via a perpetual stalemate. But in terms of technique, the stalemate is best achieved by knocking all the opponent's pieces from the board before the start of the game and then preventing the opponent from putting them back on. The defender fears or suspects that if he actually lets the game proceed, he will face defeat. In this scenario, the better chess player cannot win because she cannot play her pieces.

Despite their defensive essence, anti-access and area-denial strategies are not exclusive to a defending or status-quo power or to a tactically weaker opponent. For example, Imperial Japan's war against the United States in the Pacific from 1941 to 1945 was an anti-access strategy designed to preserve aggressive conquests achieved in Asia starting with Japan's capture of Manchuria in 1931 and continued in the wake of the Pearl Harbor attack. By the time American forces could respond effectively, Japan had already consolidated the core of its wartime empire and its conquest of the oil-rich Dutch East Indies was under way. Its objective was now to prevent reentry to the western Pacific by the one opponent with the potential for strategic superiority. With elimination of U.S. forces from the Philippines, along with defeat of the British and Dutch, the Japanese were now the tactically superior force in East Asia and the western Pacific. Their strike at Pearl Harbor was not intended as a prelude to an invasion of Hawaii or the continental United States, but to knock over the chessboard so that the Americans would decide that—as far as the Asia-Pacific region was concerned—it was too costly to put their pieces back into the game. This constitutes a classic anti-access approach.

Another metaphor useful for illuminating the concepts of anti-access and area denial—and, more specifically, describing how they have been historically overturned—is to discuss the construction of *great walls* and attempts to breach them. This perspective is of anti-access as an element of grand strategy. For the purposes of this study, great walls are meant to indicate something other than defensive structures such as city walls, fortified positions, or defenses surrounding an armed camp. Great walls herein are interconnected series of defensive—and often offensive—capabilities designed to prevent the penetration of forces or ideas that might challenge attempts to consolidate or maintain internal control. To use modern terms, we can describe great walls as anti-access networks. When the first Chinese emperor initially started the construction of the Great Wall and the Romans began Hadrian's Wall in Britain or comparable fortifications in Gaul or Germany (such as the Limes Germanicus), they were not intended to represent the limits of consolidated power or to delineate or defend ethnic divisions. In both cases, these empires had yet to consolidate absolute power within the limits of the walls. In both cases, many of the peoples living closest to one side of the wall were ethnically similar to those living on the other side. Rather, the constructions were primarily intended to prevent hostile forces outside the walls from supporting or assisting potentially hostile forces within the walls in challenging the central authority. The goals were both to deny access to the empires by forces that could destabilize the empires *and* thus allow for a consolidation of imperial authority within the walls.

This is more obvious in the case of Rome, which had almost as many "barbarians" within its great walls as it had outside. When Rome fell as an empire, it was not because outside forces had breached the walls as much as it was because those barbarians inside—admittedly with pressure or support from outside—no longer viewed loyalty or acquiescence to Rome as in their interest. This is another illustration of how an anti-access–focused security strategy—the construction of great walls—was an attempt to prevent interference with the consolidation of conquests, albeit over a much longer period of time than the efforts of Imperial Japan. It is also an illustration of how anti-access strategies have often been defeated in history: the forces maintaining the pressure outside the great walls provided the strategic opportunity for forces within the walls to combine with them to break the central authority. Arguably, this was how the communist Iron Curtain was broken in Europe. In this case, the great walls—the Berlin Wall being the most tangible—were more specifically designed to consolidate authority and keep the people within the empire than to prevent the attack of military forces from without, an unlikely prospect given Soviet military strength.

As can be sensed from the previous discussion, this book is not intended to be a detailed assessment of current technologies or weapon systems that may be used as components or counters of anti-access or area-denial strategies. Nor is it intended to be an argument for or critique of the AirSea Battle concept that is now a prominent feature of current U.S. defense policies. These topics are indeed discussed in the following pages but not with the depth they fully deserve. Since they are the focus of writing and debate by other national security professionals, I will leave the detailed examination of the more current strategic challenges to them and have concentrated in this book on the principal elements of anti-access and area denial and the fundamental requirements necessary for the defeat of such strategies. Specific scenarios have been and will continue to be the focus of war games—official and unofficial, public and secret—as well as media speculation and scholarly assessment. States identified as pursuing anti-access and area-denial strategies include the Islamic Republic of Iran and the People's Republic of China, both of which routinely express a varying degree of hostility toward the United States and its allies. Although some of these scenarios are addressed in this book, my primary attention remains on the strategic principles and history of anti-access and area denial because these are the areas that have been publicly underexamined. I know of no other book that scrutinizes them.

At this point, it is important that I explain a little about how the terms anti-access and area denial are used in this book. With the publication of the Joint Operational Access Concept by the U.S. Department of Defense (DOD) in November 2011 and its formal signature by the chairman of the Joint Chiefs of Staff in January 2012, the terms have acquired nearly official DOD definitions, which will be discussed in later chapters. These definitions certainly have their virtues in helping to codify official doctrine. They also provide a distinction between what the DOD considers anti-access operations and what it considers area-denial operations. However, my initial encounter with the concept behind the terms dates back to 1991, long before they were popularized or anyone tried to attach rigorous definitions. I have watched their meanings gradually change as a consequence of DOD internal politics and service rivalries, an evolution that will be discussed later. My view is that the attempt to codify rigorous distinctions has inadvertently washed away some of the richness contained in the overall concept.

For example, anti-access operations as a component of grand strategy naturally include international diplomatic, political, and economic activities, none of which fits well under the current set of near-official definitions. Likewise I would argue that underlying both the terms anti-access and area denial is but one essential concept, with the nuanced distinction between them being the result of the U.S. Air Force staff (and later the U.S. Army staff) refusing to utilize a term deemed too naval in its origin. Recognition of the somewhat tenuous nature of the distinction has resulted in the frequent use of anti-access/area denial as a composite term. Recent use of the acronym A2/AD—credit for which must go to the Center for Strategic and Budgetary Assessments—completes this recognition. For the purposes of this book, however, I will gradually narrow my usage to the sole term anti-access in referring to the joint, underlying concept. This is intended to avoid continual use of the awkward composite anti-access/area denial and the somewhat jarring A2/AD. My experience is that many readers outside of the DOD are not so enamored of the military predilection for space-saving acronyms, and—facetious or not—A2/AD conjures up images of such popular fictional icons as R2-D2 and C-3PO. Hence, anti-access will be the term I use most.

Chapter 1

A Tale of Two Wars

This book begins with a tale of two wars, one in 480 BC and one in 1991. The first war is the earliest example of an attempted anti-access strategy for which we have a substantial historical record. The second generated the modern interest in anti-access and area-denial operations, particularly by regimes that could envision possible hostilities with the United States and its allies and partners.

Overwhelming Odds

In the spring of 480 BC, the independent city-states of mainland Greece—including the nascent democracy of Athens—faced a grave threat to their existence. The largest armed force ever before assembled, led by the Persian emperor Xerxes, self-proclaimed "king of every country and every language, the king of the entire earth," had begun its campaign of conquest and revenge.[1] Herodotus, known as the "father of history" for the validity of his professional inquiries, records that Xerxes' army numbered over 1.7 million troops, supported by 1,327 warships. Modern historians dispute the first figure as being far beyond the logistical support possible in the ancient age, but some concur that he may have arrived in Europe with as many as 150,000 to 200,000 warriors, along with the inevitable, innumerable camp followers.[2] In contrast, the average Greek city-state—jealous of sovereignty and rarely before united—had but a few thousand defenders. The larger cities, such as Athens, Corinth, and Sparta, might be able to muster eight to ten thousand infantry hoplites. Clearly, they faced an overwhelming foe.

Even before Xerxes' force crossed the Hellespont—the narrow strait that separates the Black Sea from the Mediterranean—on a bridge of nested boats, the northernmost Greek cities began their capitulation. Those relatively few that refused—located in central Greece, the islands, and the large southern peninsula called the Peloponnesus—united in an alliance led by the Spartans and Athenians. Yet even when unified, the Greek forces were vastly outnumbered. They were well aware that if the Persian forces were able to operate south of sacred Mount Olympus and the other mountains of Thessaly and lay siege to individual cities, none could survive. Positioning the Greek forces to utilize geography in denying Persian access to the populated region appeared to be the only possible strategy.

Plutarch, a Greek moralist and historian writing about five hundred years later, credits the Athenian navalist Themistocles, who had been elected *strategos* (combination admiral and general), with first promoting the anti-access approach. In Plutarch's words, "having taken upon himself the command of the Athenian forces, he [Themistocles] immediately endeavoured to persuade the citizens [referring to males of military age] to leave the city, and to embark upon their galleys, and meet the Persians at a great distance from Greece; but many being against this, he led a large force, together with the Lacedaemonians [Spartans], into Tempe, that in this pass they might maintain the safety of Thessaly, which had not yet declared for the king."[3]

Themistocles' original intent may have been to sail the Athenian galleys to destroy the bridge of boats across the Hellespont; we know that he publicly advocated such a plan following the later battle of Salamis.[4] Such an action would indeed have met the Persians "a great distance from Greece" and conceivably prevented the movement of land forces between Asia and Europe. But this was not a strategy his countrymen were used to, and they undoubtedly viewed it as extremely risky. Instead, he initially had to be content to attempt an anti-access defense by land.

Stands on Land and Sea

The Vale of Tempe is a gorge approximately 6.2 miles (10 kilometers) in length through which the Pineios River flows between Mount Olympus and Mount Ossa before it reaches the sea. It narrows in some spots to 82 feet (25 meters), with cliffs rising to 1,640 feet (500 meters). (Today it is infamous for fatal car accidents.) In ancient days it was considered the haunt of Apollo and the Muses, and at a distance of 146 miles (235 kilometers), it would have been a fourteen-day march north from Athens.[5]

Themistocles and the other Greek commanders apparently considered the Vale of Tempe to be the most defensible physical feature at which their much smaller force could stop the army of Xerxes. But that was based on inexact geographic knowledge or what today might be called incorrect geospatial intelligence.[6] Once in position, the Greeks were warned by the King of Macedon—who had a treaty with Xerxes but was still friendly to the city-states—that there were a number of other open passes to the west by which Xerxes' army could flank their position. At the same time, through human intelligence it became known that Thessaly (south of the Vale) intended to capitulate. Deeming his own force insufficient and convinced that a maritime-centered defense would be best, Themistocles returned the force to Athens—as Plutarch somewhat uncharitably describes—"without performing anything."[7]

However, the retreat did not shake the overall Greek commitment to an anti-access approach. Another narrow pass lay to the south about 62 miles (100 kilometers) below the Vale of Tempe and 85 miles (136 kilometers) north of Athens, at a spot where the mountains ran to the water's edge at the Sound of Euripos between the large island of Euboea and the Greek mainland. This was Thermopylae—to be the site of the last stand of the three hundred Spartans under King Leonides, an act of bravery heralded in poetry, prose, and most recently film. Owing to almost 2,500 years of silting and other geological changes, the location is no longer the mere ribbon of land where three hundred or so troops could delay tens of thousands. But during the war with the Persians, this area of hot springs and geothermal activity was another favorable site for the Greek anti-access defense.

This time the land force of approximately ten thousand hoplites was led by the Spartan Leonides, with Themistocles commanding a Greek fleet of 333 warships, based at Artemisium on Euboea to prevent Xerxes' ships from landing additional troops and to impede their transit south through the sheltered waters of the sound. The combined army-navy effort reflected the geographic reality of all anti-access operations, and whether or not Themistocles was the very first to grasp this reality, his effort to persuade the other Greek leaders to prioritize maritime operations was forceful and unwavering. Like Leonides, Themistocles faced a superior enemy: the Persian fleet had suffered heavily from storms but still numbered above nine hundred ships. But unlike the Spartan king, he retained the mobility inherent in naval forces that prevented them from being flanked in narrow seas. In three engagements, the Greek fleet was able to fend off Persian attacks at sea, while simultaneously launching a propaganda campaign to get Xerxes' Ionian Greek vassals—who constituted some of his best sailors—to defect.

As is well known, the anti-access operations at Thermopylae-Artemisium failed through treachery when Xerxes found a Greek scout with knowledge of a shepherd's path through the mountains. Treachery and defections were routine "game changers" in ancient wars. Recognizing that he was being flanked, Leonides retained a force of three hundred Spartans and about a thousand other Greek troops while ordering the rest to retreat. Whether Leonides intended a last stand or a rearguard action, his reputation for utmost bravery was made eternal when the Spartans fought with their spears until they were broken, fought with their swords until they too were broken, and then fought with their hands and teeth until they were all dead.[8] The Greeks lost about four thousand warriors at Thermopylae, the Persians about twenty thousand. Xerxes was so enraged by the losses and delay that after locating the body of Leonides, he cut off the head and stuck it on a pole, an act then considered particularly insulting for the body of a king.[9]

Closing the SLOCs

At this point it appeared that the Greek alliance would unravel and Xerxes would inevitably conquer the whole of Greece. While the city-states of the Peloponnesus, notably Sparta and Corinth, had yet another narrow land barrier—the Isthmus of Corinth—that could be fortified to delay Xerxes' march toward their cities, Athens and others were now open to destruction.[10] Their choices included surrender or flight to exile in the Peloponnesus or possibly to Greek-controlled Sicily. Here again Themistocles rose to the fore, utilizing the Delphic oracle of the "wooden walls," alliances with political enemies, and his powerful rhetoric in the Athenian assembly to persuade his fellow citizens to abandon the city but fight on in the ships. He used threats of defection, bribes, faked treachery, and a fait accompli to force the other Greeks to fight the famed Battle of Salamis instead of fleeing in their ships back to the Peloponnesus.[11]

The Battle of Salamis—fought in the narrow channel between the island of Salamis and Athenian territory—is often cited as an example of how sea power can be used to ultimately defeat forces on land, as well as how a smaller naval force might use stratagem and coastal terrain to destroy a larger fleet.[12] The critical importance of morale and patriotism can be invoked too; the Greeks were a voluntary alliance of politically equal city-states, whereas the Persian navy consisted of ships from conquered peoples, many—like the Ionian Greeks of Asia—presumably serving unwillingly. But in addition, the logic behind Themistocles' plan can be cited for his understanding that the sea was as important as land terrain in ultimately denying Xerxes the logistics necessary for military action. Land terrain may have failed the Greek anti-access attempt, but the control of the sea meant control of the transregional supply lines necessary for feeding Persian forces.

For ancient armies, Napoleon's later aphorism that "an army travels on its stomach" was very apt. Starvation was a routine planning factor. So was the availability of water.[13] The tyranny of distance mandated routine supplies by sea, particularly in a world where beast-drawn wagon trains could travel at best 10 miles (16 kilometers) per day, but a fortunate breeze could move a supply ship over 120 miles (193 kilometers) in the same twenty-four hours. The maritime dominance achieved through the Battle of Salamis meant that the Greeks could control the sea lines of communications—or in standard abbreviation, SLOCs—from Asia Minor to Europe, denying access to the logistical support essential for maintaining Xerxes' army.

In original usage, the *communications* part of the term SLOCs does not merely mean messaging or information flow, but—as derived from corresponding military usage in nineteenth-century land warfare—includes the ability of divided forces to recombine, as well as be resupplied or reinforced. Today, with most information traveling electronically, certain analysts and scholars have suggested that the concept of SLOCs might be antiquated. But that view fails to recognize that the term is intended to include the physical transfer of matériel and personnel. Perhaps "sea lines (or lanes) of commerce" is a more appropriate modern depiction.

In any event, control of the SLOCs turned the sea into an anti-access barrier for Xerxes, even more limiting than any narrow pass on land. The situation that faced him was a logistical nightmare. The city of Athens had been destroyed, but Xerxes' force was simply too huge to live off the land, much of which lay denuded. His troops could not survive without access to grain, and they had previously eaten all that could be stripped from the conquered northern city-states. An extensive supply of grain could only come by sea, and now the Greeks controlled the sea.

As noted, Themistocles' instinctive reaction following Salamis was for the Greek fleet to immediately sail to the Hellespont and destroy the bridge of boats. That would ensure Persian supplies could move *neither* on land nor sea. But as his fellow strategists were quick to point out, it was more advantageous—at that point in time—to allow Xerxes the opportunity to quit the region and retreat into Asia than to trap him in Greece. Ever the schemer, Themistocles had to be content with sending a false message to the Persians claiming that, in order to allow Xerxes to return his royal personage to Persia, he himself was restraining the other Greeks from racing to the Hellespont.

Whether or not this message had an effect, Xerxes ordered the bulk of his forces—already starting to face starvation—back into Asia. There, a number of revolts that required his repressive hand had meanwhile been brewing. The cost of conquest of all of Greece had suddenly become too high. The twenty thousand or so troops he left behind were destroyed in battle in the following year by the Spartan-led Greek land army.[14] It is doubtful whether Xerxes actually believed that the forces he left behind would survive, particularly after experiencing Greek determination and ferocity at Thermopylae.

It was thus the naval victory scripted by Themistocles that denied the region to Xerxes, defining regional anti-access operations as a joint and combined effort, with a strong—generally dominant—maritime component. For those who love historic irony, Themistocles was later exiled from Athens and forced to flee to the Persian court. But that's another story.

Five Fundamental Elements

An evaluation of the war between the Greeks and the Persians introduces five fundamental elements that combine to sketch the outlines of the concept of anti-access. Other factors may have contributed to the success or failure of any particular anti-access defense or counter–anti-access campaign, but later case studies will indicate that these five elements are indeed common to the construct of anti-access and area-denial strategies across history.

The five fundamental elements can be summarized as:

1. The perception of the strategic superiority of the attacking force
2. The primacy of geography as the element that most influences time and facilitates attrition of the enemy
3. The general predominance of the maritime domain as conflict space
4. The criticality of information and intelligence, and—conversely—the decisive effects of operational deception
5. The determinative impact of extrinsic events or unrelated events in other regions

These elements can be analyzed independently, but they are not truly independent factors. Rather, they function together in determining the strategic environment in such a way that adoption of an anti-access defense posture becomes a logical strategic choice. In that sense, they can be viewed as *defining* and *determining* factors for both decision-making and outcome.

In brief, without a perception that the opponent is strategically superior, optimizing one's military resources primarily for an anti-access approach is not necessarily an appealing choice. Instead, one's own strength on the battlefield and capacity to operate out of area can even more effectively deter any possible attack. An anti-access strategy costs resources that could be utilized to optimize one's battlefield strength. Without favorable geography, it becomes difficult to channel and thus reduce the options of a strategically superior opponent. Since it is easier to move overwhelming military force by sea than any other medium, the maritime domain inevitably becomes the decisive conflict space in any anti-access versus counter–anti-access campaign. Without adequate information and intelligence, the anti-access force—limited by its inferiority in strength—cannot determine the most suitable locations to deploy so as to meet the invader's main thrust. Likewise, without adequate information, the counter–anti-access invading force cannot determine which access route best bypasses the strongest defenses. Effective deception thereby becomes a premium on both sides. Meanwhile, events outside of the region—such as the possibility of rebellions in the Persian Empire—maintain a constant pressure on the choices of the from-out-of-area invading force. In determining whether to adopt an anti-access approach, one must assess as a critical factor the ability to influence extrinsic events in such a way as to distract the strategically superior force or to change its decision-making. Diplomacy, economic relations, other international political activities, and overt or covert military support can be factors in influencing extrinsic events, which is why anti-access strategies naturally involve more than military operations.

Strategic Superiority of the Attacking Force

The first common element is the perception of the strategic superiority of the attacking force. It is this perceived asymmetry in strategic force that motivates the defender to specifically focus its resources on denying regional or area access as its primary operational effort. Without this perception, anti-access efforts might be viewed as productive elements of an overall defensive campaign but would not necessarily be considered to be the most critical components. If an attacker could be defeated in a force-on-force engagement within the contested region, preventing the attacker from entering the region would be an operational luxury or adjunct, not a necessity around which the entire defensive campaign must be built. The potential for force-on-force victory by the defender could also function as an adequate deterrent.

Strategic superiority does not necessarily mean technological superiority, but in general technological superiority is a component of the former. Many studies of anti-access have emphasized the dominance of high-technology weapons, and in modern scenarios they are certainly the key operational concern. As will be seen, technological advantage has been a significant factor in a number of historical examples of anti-access operations. However, technological advantages have appeared on both sides—attacker and defender—and the diffusion of technology has often meant that any specific advantage has not lasted for long. The strategically inferior force has often attempted to neutralize technological advantages of the strategically superior force by stratagems, tactical innovations, or unexpected uses of current technologies—what we would refer to today as asymmetrical warfare.

In the case of the Athenians, there was little practical debate about whether they could actually defend the city—barring divine intervention—from a surrounding Persian force. In truth, there was quite a debate about the possibility of divine intervention, something all Greeks considered a routine feature of momentous events. This debate was not resolved for all; a small party convinced that the prophecy of the "wooden walls" meant a wooden palisade around the Acropolis stayed in the city to be slaughtered by the Persians even as the rest of the citizens were manning the ships at Salamis. Themistocles had convinced most that "wooden walls" was a divinely inspired metaphor for the warships. But in any event, the superiority of Persian forces in numbers and the inevitability of Greek defeat in a land war of attrition were never seriously questioned.

Thus it was their perception of Persian strategic superiority that drove the Greek strategists to investigate the possibility of an anti-access approach. This was different from the then-common practice of attempting to wait out sieges until the attackers got tired and a less onerous agreement was negotiated. Even the city-states with extensive walls could not withstand a siege by such a vast army. Seeking decisive battle on flat home ground was likewise out of the question. In terms of technology, there was no apparent advantage on either side.[15] The Greeks, particularly the Spartans, may have been excessively brave, but they also wanted to win. The only possibility of victory lay in denying Xerxes the chance at striking the cities themselves. That the strategy initially failed—Athens was after all burned to the ground and its lands despoiled—points to the inherent difficulty of conducting a successful anti-access campaign against a truly determined opponent. The operational objective remained the neutralization of the superior force until time, attrition, and/or extrinsic events shook the determination of the attacker.

Primacy of Geography

The second fundamental is the primacy of geography as the element that most influences time and facilitates attrition. Admittedly, anti-access efforts are dependent on more than geographic characteristics. However, geographic characteristics are the most permanent factors that can be utilized in denying unfettered operations within a contested region.[16] The technologies of war and the political divisions of the world may change, but geography does not (or does so very slowly).[17] Mountainous terrain, narrow passes, isthmuses or straits, littoral features such as islands, bays, shallow waters, and climactic differences are primary geographic discontinuities that strengthen potential defenses, whether in local, regional, or global conflict. Those land areas that have natural barriers such as mountain ranges (the territory of the Swiss Confederation, for example) have had less of a historical experience with successful enemy invasions than territories bordering flat plains (Poland being an opposite example).[18] The sea, too, has proven to be a barrier—perhaps the most significant barrier—when utilized effectively. It became the primary anti-access barrier in the war between the mainland Greeks and the Persian Empire.

It would be incorrect, however, to imply that specific geographic features are absolutely insurmountable barriers. In the era of air and space travel, there are few features that can provide protection from air attack and few, if any, that cannot be overflown. But the reality remains that geographic conditions still continue to limit the type, direction, and scale of military operations. Lighter forces can be transported by air over geographic barriers, but substantial amounts of heavy equipment cannot, or at least not without tremendous cost. In contrast, heavy equipment can be transferred across the sea at consistent speeds that cannot be achieved on difficult land terrain. This points to the fact that in developing a regional anti-access strategy, effective planners must take into consideration the totality of geographic features—not merely in determining the positioning of defending forces, but also in determining the type of weaponry and force structure itself in which the defender should invest.

The need to tailor forces to the geography also points to the fact that modern counter–anti-access operations need not be the sole province of naval and air forces. Ultimately the goal of the counter–anti-access force is to strike at the vital center of the enemy and cause his capitulation, an outcome that generally requires land forces to achieve and enforce.

Like all aspects of an integrated campaign, counter–anti-access operations are combined arms warfare—and in current terminology a cross-domain function. But there is a caveat to that: geographic and other planning considerations mandate far different quantities of the distinct forces involved. Some domains or mediums may require more extensive forces than others.

Developing an anti-access strategy requires optimizing the defending force to best use the permanent geographic features in order to increase the costs to an invading force, additional costs that would not occur but for existence of the geographic features. Resources, including research and development, may have to be allocated by the anti-access force so as to extend the effects of existing features. From the opposite perspective, an attacking force determined to counter an anti-access strategy must optimize its own force to overcome the effects of any potential geographic barrier. But countering barriers always entails deliberate planning and resources that may necessarily (or at least deliberately) be spent elsewhere. The Persians could have built a more effective, unified navy, but it would have cost greater resources in leadership, time, and expertise than they chose to pay. The Greeks used their more natural, relatively more extensive seafaring tradition, which was reflected in their political organization, to take control of the dominant geographic feature of the eastern Mediterranean. Themistocles' insistence on pouring Athenian resources into naval forces reflected this perception that maritime geography provided the greatest anti-access leverage.

Predominance of the Maritime Domain

That the Greek strategy did not completely fail points to a third fundamental element, one that flows naturally from the primacy of geography: the predominance of the maritime domain as conflict space.[19] An elementary fact of global geography is that over 70 percent of the planet is ocean. Inevitably a military operation that extends from one global region to another requires the attacking force to transit maritime regions.

While maritime regions have their own geographic characteristics, a significant portion of their maneuver space most closely resembles the open plains that favor the maneuver of vast armies and aviation assets. In that sense, it negates defensible geographic features, although it is a maneuver space requiring specialized platforms, including those that can travel submerged. The ability to utilize maritime regions is therefore the most significant advantage that an interregional attacking force can possess, and conversely the ability to deny an attacker's use of maritime regions is a dominant factor in the success of any anti-access campaign.

Since the modern concept of anti-access was derived from naval roots, it may seem easy to dismiss the third element as tautological. The separations between geographic regions usually include bodies of water, therefore maritime space becomes important in any anti-access or counter–anti-access campaign. But this misses the recognition that control over the maritime region can act as a surrogate for control over adjacent land. This is the premise for the concept of sea basing, which is discussed later. Right now, it is sufficient to note that in the case of the Greek-Persian War example, Greek efforts to use geographic land barriers to keep the Persian force from their cities could only be successful if the Greeks also ensured that Persian maritime forces could *not* be used to bypass the land barriers by landing amphibious forces beyond them. Persian efforts to overcome the Greek land-based anti-access could only be successful if the Persian force *could* retain at least partial use of the sea to resupply its overwhelming and resource-intensive land force. This was vastly different than the situation faced by the Persians in conquering and maintaining their empire in the Near East and Asia Minor, where maritime conflict seemed peripheral. To overcome the Greek anti-access defenses and conquer Greece, maritime conflict became central.

When the campaign is viewed from the joint paradigm of twenty-first-century strategy, emphasizing the centrality of the maritime domain in major combat operations seems almost gauche. Two factors are in play here. First—and protestations of advocates aside—the concept of *jointness* now carries the connotation of the equal importance of all combat domains and service organizations. To emphasize the dominance of one is perceived as marginalizing the others and therefore neither cooperative nor orthodox. Secondly, operational domains are now being defined for convenience along particular service lines, which leads to a growing disconnect in strategic thought. Understanding the *predominance of the maritime domain as conflict space* in fashioning or countering any anti-access strategy requires one to push aside current conventions and return to the view that the maritime "domain" includes the air space above the oceans and littorals as well as the water itself. That is why modern navies consist of platforms operating under the sea, on its surface, in the air above, in littoral regions, and in space and cyberspace.

Criticality of Information and Intelligence

A fourth fundamental element is the criticality of information and intelligence. This is true of all warfare, but it is particularly true in tailoring a defense designed to prevent strategically superior forces outside of one's immediate region from entering. The anti-access force must seek to deny information to the enemy and deceive enemy forces. Conversely, operational deception is a premium counterweight in the hands of the strategically superior force. The Greeks' attempt at a defense at Tempe failed because they did not have an accurate picture of the geography and had made invalid assumptions. In contrast, they had a much clearer picture at Thermopylae and at Salamis (particularly the Athenians), which allowed them to engage the enemy more effectively.

The Greeks also had an accurate knowledge throughout the conflict of the size, composition, disposition, and psychology or morale of Xerxes' forces. This was not a testament to Greek intelligence-gathering; rather, it was because Xerxes continuously broadcast it to them. All the ancient sources maintain that when Xerxes captured Greek spies, he let them go so they could return home to tell all about the size and power of his army. In that way Xerxes thought that the Greeks would be more likely to surrender before he arrived. This proved to be only partly effective and was based on intelligence assessments that included mirror-imaging, viewed in modern intelligence as a major flaw. Apparently it was not until the slaughter of his forces at Thermopylae that Xerxes called forth the Greek defectors in his train to inquire seriously about the psychology and motivation of the Spartans.[20]

Xerxes may not have utilized confounding techniques of deception, but the Greeks did. Themistocles' genius in this misinformation effort lay in exaggerating the truth about the disunity among the Greek leaders and in reinforcing Xerxes' perception that getting a prominent Greek leader to defect with his forces was only a matter of time. This confirmed what the Persians wanted and expected to hear.[21] Thus, when the Persians encountered stubborn Greek resistance instead, they continued to be tactically surprised.

Projecting to anti-access strategies in today's age of information-based, high-technology warfare, the *criticality of information and intelligence* seems even more obvious and a motivating factor toward the development of joint doctrinal and resource collaboration such as the AirSea Battle concept. It has also given impetus to the recent focus of cyberspace operations, although the two very different functions of cyberdefense of the computer assets of civil society and defense (and offense) of military forces are often conflated.

Determinative Impact of Extrinsic Events

A fifth historical element illustrated in the Greek-Persian War example—and even more prominent in other cases—is the determinative impact of extrinsic events (unrelated events in other regions) on the success or failure of the anti-access strategy. This would seem natural, since extrinsic events have often had an impact in modifying the calculus and motivations of combatants in any war, no matter the strategies adopted. However, in a situation where the attacker's home territory lies outside the contested region and with its political interests likely to be global or multiregional in nature, extraregional events—either orchestrated by or having nothing to do with the defender—are likely to have an even greater impact on the attacker's focused effort at countering the anti-access campaign.

Earlier it was pointed out that—at its essence—the operational objective of an anti-access strategy is the neutralization of the superior force until time, attrition, and/or extrinsic events shake the determination of the attacker. Identifying such extrinsic events as the fifth element is intended to emphasize that anti-access strategies are adopted in situations of asymmetry in terms of military power, objectives, and motives, but that outside events can impact this asymmetry—potentially equalizing relative military power.

In the case of the Greeks' anti-access efforts, their gaining control of the SLOCs necessary to feed Xerxes' army occurred in the context of perceived instability within the Persian Empire itself. The defeat of the Persian-controlled fleet did not mean the end of combat in Greece. It took a land battle at Plataea, approximately forty miles (sixty-four kilometers) west-northwest of Athens, in the following year to eliminate Persian

TABLE 1-1 Fundamental Elements of the Greek-Persian War

Fundamental Elements	Greek-Persian War 481–478 BC
Perception of strategic superiority of attacking force	• Greeks recognize overwhelming odds • Persians broadcast superior numbers in effort to encourage capitulation of cities • Greeks conduct operations to keep superior Persian forces away from their cities
Primacy of geography	• Themistocles proposes cutting bridge of boats across Hellespont linking Asia to Europe • Greeks attempt defenses in narrow passes and narrow seas • Greek navy defeats large Persian naval force in narrow strait between Salamis and mainland
Predominance of the maritime domain	• Persian forces dependent on resupply from sea • Greek land defenses dependent on naval protection against flanking operations • Greeks secure sea control over the Aegean, denying effective resupply to Persian forces in Greece • Greeks carry war back to Asia Minor through maritime operations, attempt to control ports of embarkation
Criticality of information and intelligence	• Xerxes broadcasts Persian force size, disposition, and objectives • Xerxes rejects reports on Greek morale, commitment until battle at Thermopylae • Knowledge of flanking routes critical to success or defeat of defense in narrow passes • Themistocles utilizes deception to force Persian decisions
Determinative impact of extrinsic events	• Potential for revolts within Persian empire provide cause for continuing concerns and affects Persian decision-making • Greeks attempt to encourage revolts

land forces from the Greek mainland. On the same day, the remnants of the Persian fleet were destroyed on the beach near Mount Mycale on the opposite side of the Aegean Sea in Ionia (modern Turkey).[22] Following these Persian defeats, Greek fleets continued the war by assisting the very revolts in Ionia and, later, Egypt that Xerxes had feared. The extrinsic events that may have played a role in Xerxes' calculations to quit his main effort in mainland Greece came to pass in an ironic fashion. The Persian Empire itself did not fall as a direct result, but its focus on internal conflicts kept it from returning across the Aegean.[23]

The Greeks had won their anti-access campaign and were determined to prevent any possibility of a Persian return by wrestling control of the Persian Empire ports themselves. They did this by overt and covert support to independent-minded city-states of the Asian coast, including operations on the strategic island of Cyprus and conquest of the city of Byzantion controlling the Hellespont, later to be renamed Byzantium, Constantinople, and Istanbul. The struggle for control of the ports of embarkation of the strategically superior power subsequently became a feature of other successful anti-access campaigns.

Table 1-1 summarizes and correlates activities of the Greek defense against Xerxes' campaign to the fundamental elements of anti-access operations.

The Transformational War

The Greek anti-access campaign of 480–479 BC is, of course, *ancient* history. We live in an age when many perceive technology to have so transformed warfare as to make much of the history of past warfare decidedly ancient indeed and likely irrelevant. They are wrong. But rather than argue the validity of historical relevance, let us fast-forward to a more immediate past to examine the second war—the one that made public the very concept of *military transformation*, as well as provided the justification for twenty-first-century anti-access strategies.

Today the Gulf War of 1991 seems overshadowed by subsequent, lengthier conflicts with their less conclusive results. But it was the tremendous U.S.-led coalition victory against Saddam Hussein's forces in Kuwait and southern Iraq that provided the impetus for the modern examination of anti-access and area-denial strategies, as well as the overarching concept of a "revolution in military affairs" (RMA). The techniques of anti-access and areadenial have been used throughout history, but it was not until after the Gulf War that they began to be publicly articulated as an independent concept.

Likewise, it was the systematic evaluation by the militaries of other nations of the U.S. capabilities demonstrated in Operation Desert Shield and Operation Desert Storm—primarily by those who chose not to participate in the coalition and particularly those who could conceive of future hostilities with the United States—that justified the acquisition of sensors and weapon systems optimized for the *denial* of contested area rather than the *control* of it.

One must be clear, however, not to imply that the choice between optimizing for denial versus control is a particularly new idea. Control implies the ability to dominate a combat space and utilize it for one's own operations. Denial is meant to indicate that the use of the combat space is denied to the opponent but cannot necessarily be utilized by oneself. In other words, denial indicates that neither combatant can effectively utilize the combat space for their own purposes without great risk to their forces. Naval strategists have long identified *sea denial* and *sea control* as alternative approaches to sea power, but with the assumption that opting for a sea-denial approach indicated recognition of a weakness that could eventually be remedied. Sea control—also termed command of the sea—has been traditionally assumed to be the direct objective of naval power. Sea denial was (and is often) perceived as the necessarily limited objective of a weaker fleet.

In the case of the Soviet navy—whose forces were optimized for most of their existence for a sea-denial role—there was a presumption on the part of Soviet naval leaders, notably Admiral of the Fleet Sergei G. Gorshkov, that eventually the Soviet navy would increase in capability until it could challenge the U.S. Navy for command of the seas.[24] But in contrast, what was different in the foreign analyses of the Gulf War was a growing recognition that the military technological advantage of the U.S. armed forces not only proved decisive on the battlefield, but the tactics, training, and determination of U.S. conventional forces could not be matched. *Denial* was no longer perceived as but an interim step toward building the capacity toward *control*. Following the Gulf War, it was now perceived as a goal in itself, since control appeared impossible to achieve. The results of Desert Shield/Desert Storm transformed the conception of how a conventional conflict was to be fought, particularly against the United States.

Extent of the Victory

The importance of the Desert Shield/Desert Storm victory to the subsequent evolution of modern military strategy is much out of proportion to the liberation of the small State of Kuwait or the twelve-year respite in the theater of expansionism that marked Saddam Hussein's regime. While the use of high-technology weapons is readily apparent, the transformational effects of the Gulf War in terms of strategy and operational art—an interim level of military planning and operations between strategy and tactics—have become a bit obscured by a decade-long focus on counterterrorism and counterinsurgency. It is hard to reconstruct the impact unless one reads (or rereads) such prewar reporting as appeared in the prestige media, in which it was frequently repeated that Iraq had the "fourth largest army in the world" and "sixth largest air force" with "approximately one million men under arms," was "tested in recent combat" in a brutal war against Iran, and willing to take horrific losses. As one brief history of the conflict notes, "great doubts haunted the coalition forces before the battle for Kuwait. . . . For those in the United States, the bitter memory of Vietnam hung over them like an unwelcome spectre. Critics suggested that the effect of excessive casualties would kill the campaign because American society would no longer accept such losses. Images of trench warfare that had characterized the Iran-Iraq War and, of course, the bloody stalemate of the First World War in Europe were brought back by military pundits who fed the desire of news networks for a prediction of the outcome."[25] That this was believed by Saddam Hussein himself and was a primary element of his calculations concerning war is quite evident. Saddam stated to American ambassador April Glaspie prior to the invasion of Kuwait, "Yours is a society which cannot accept 10,000 dead in one battle."[26]

Other contemporary sources pointed to an inevitable coalition victory, but estimates as to casualties and effectiveness of Iraqi forces varied. Few, if any, envisioned that the conflict would end after but thirty-eight days of air operations and four days of ground combat, with only 246 coalition personnel (147 Americans) killed in action as opposed to a conservative estimate of twenty thousand Iraqi troops killed and over sixty thousand captured.[27] While air operations over Baghdad—with the use of stealth aircraft, Tomahawk cruise missiles, precision ordnance, and a host of electronic warfare measures—reduced the command-and-control capabilities of the extremely centralized Iraqi force and captured public imagination, the effectiveness of coalition ground and naval forces also far exceeded historical standards for conflicts including forces of such size. In terms of military history, the size of the victory was stunning, entailing not just the ejection of Iraqi forces from Kuwait itself, but in essence the capture of all Iraqi territory south of the Euphrates River after about one hundred hours of ground combat against a numerically equal or even superior force. Coalition forces anticipated and defeated every effort of an Iraqi military considered to be one of the most powerful in the region and possessing extensive Soviet military resources. For other militaries, particularly those also built along the Soviet model, it was a great shock. At the beginning of the war the Iraqi order of battle was credited with some 5,500 main battle tanks, 1,500 armored fighting vehicles (AFVs), and 3,500 pieces of field artillery. Official DOD estimates state that 3,847 Iraqi tanks, 1,450 armored fighting vehicles, and 2,917 artillery pieces were destroyed by air and land attacks during the total forty-three days of combat operations.[28] In contrast, coalition losses were miniscule, with more the result of unfortunate blue-on-blue engagements than from Iraqi fire.

Not only did the campaign demonstrate that the high-technology weaponry in which the United States had heavily invested during the Cold War proved effective, but that the operational art and tactics of U.S. forces—built around the AirLand Battle doctrine designed to stop the movement of multiple Soviet echelons in a war in Central Europe—proved far superior to standard defensive operations. Whether they would actually have defeated Soviet forces in a war between NATO and the Warsaw Pact remains mercifully unknown given the presumably much higher level of training and readiness of Soviet troops in comparison to the Iraqis. Yet few potential opponents of the United States could boast forces comparable to the Red Army, so the effectiveness of an AirLand Battle approach against their own defenses could not be discounted.

Likewise, U.S. intelligence collection by reconnaissance, surveillance, and other technical means—including the comprehensive use of space systems for tactical operations—and its effective analysis, distribution, and practical application far exceeded any comparable opposing capability, including that of the Soviet Union, then nearing the point of implosion.[29]

The courage and discipline of the coalition service members was made obvious by the Gulf War, which was an apparent surprise for the leaders of states hostile to the United States and its allies who viewed Western militaries from the prism of the U.S. "defeat" in the "people's wars" in Indochina, previously identified as the "spectre of Vietnam." While foreign professional military analysts may have had a clearer understanding of the morale and steadfastness of American forces, the perception of a lack of will on the part of Western commanders, an antiwar attitude on the part of the public, and a dubious commitment of the troops had been a routine part of the official view of the core Marxist states and their clients. Although the United States and its allies had carried out quite a number of relatively small military interventions—accompanied by the media controversies that have been a tradition in American history but seem so confusing to nondemocratic outsiders—foreign critics had discounted these successes (while magnifying the few failures) as simply the consequence of numerically overwhelming force. On the surface, the situation in the Gulf seemed much more equal, at least in terms of ground combat forces. A pre–Desert Storm quote from an Iraqi Popular Army "volunteer," likely scripted by the Iraqi information ministry for the Western media, captured this official attitude of doubt: "We [Iraq] fought an eight-year war with Iran, my dear. You only sent your army to Europe, to Grenada, and to Panama. You Americans do not know what a modern war is."[30]

A search of academic archives of war statistics, such as the Correlates of War project and the work of Lewis Fry Richardson, indicates the extent of the Desert Shield/Desert Storm *military* victory: never before in recorded history were so many enemy forces killed or captured, and so much territory occupied, with so few friendly casualties in such short a time.[31] The U.S.-led coalition seemed to know what a modern war was.

From a defender's perspective, the primary lesson learned from the conflict was the apparent invincibility of U.S. military forces in a conventional force-on-force conflict. As the DOD *Joint Operational Access Concept* document states from the vantage point of 2012, "events of recent decades have demonstrated the decisive results U.S. joint forces can achieve when allowed to flow combat power into an operational area unimpeded."[32] This created a dilemma for authoritarian states opposed to U.S. interests—particularly those viewed as pariahs by the international community or, in the words of President George W. Bush, members of the "axis of evil."

The Dog That Didn't Bark

International assessments of the inability of Saddam Hussein to mount even a credible defense against coalition forces initially—and perhaps somewhat sardonically—centered on his lack of a nuclear deterrent. This perspective is summarized best in a quote—possibly apocryphal—by former Indian Army chief of staff General Krishnaswamy Sundjari, who suggested that the real lesson was not to fight the United States unless or until one had acquired nuclear weapons.[33] Since nuclear weapons could conceivably be considered the ultimate anti-access tool, capable of devastating forces stationed at regional bases, confined to narrow straits or passages, or at ports of entry or embarkation, his assessment makes sense. Nuclear weapons as tools of anti-access will be discussed later. However, strategists have long considered the use of nuclear weapons to be a "firebreak" across which their use would result in a similar nuclear response—not a decision to be taken when facing a nuclear superpower with a vastly superior arsenal.

Rather, more detailed reactions led by military commentators in the then Soviet Union and the People's Republic of China focused on Saddam's decision *not* to challenge the unfettered entry of massive U.S. and coalition forces into the Persian Gulf region using his conventional forces. In this, the clues of his defeat are derived from what was *not* done—in a similar fashion to the fictional Sherlock Holmes' observation in one of his adventures of the "curious actions" of a dog in the night at the scene of a purported crime. When a police inspector points out that the dog had done nothing, Holmes replies that that was what was indeed curious.[34] Contemporary assessments of Iraqi strategy found it indeed curious that Saddam's forces allowed the United States almost six months to build up the striking power of two complete army corps in Saudi Arabia without taking meaningful actions to prevent it.

Contemporary Assessments

A recurring theme in Soviet assessments of 1991 was the Iraqi "failure to launch a preemptive strike against the allied coalition when the opportunity was available."[35] While details for such strikes were not provided, Soviet military doctrine placed emphasis on strikes on the enemy's rear and logistical capabilities, something Saddam's forces did not attempt during the initial buildup stages of Operation Desert Shield. In their own internal assessments, China's People's Liberation Army (PLA) commentators were a bit more specific. As an American interpreter notes:

> In addition to Iraq's economic weakness and its lack of a nuclear deterrent, Chinese analysts criticize Iraq for:
>
> - Not making surprise attacks on U.S. airbases and the U.S. rear
> - Permitting the United States time to build up its logistics and conduct special training for several months before the war
> - Not employing "special measures," such as harassing attacks[36]

Analysis of Desert Storm among Chinese military planners was particularly extensive because it occurred during a major change in their previous strategy focused on a war against the Soviet Union—a war that might involve nuclear weapons—to a focus on local, limited wars on the periphery of Chinese territory. One source notes: "In 1991, as the PLA was undergoing extensive reform and reorganization, the Persian Gulf War erupted. In many ways, the Gulf War was precisely the kind of military conflict PLA analysts had been assessing [during the change in strategy]."[37]

Meanwhile, the official post–Desert Storm DOD report to Congress, *The Conduct of the Persian Gulf War*, provides a virtual recipe for operations suitable for an anti-access campaign *not* taken by Iraq: "There was no submarine threat. Ships did not face significant anti-surface action. We had little fear that our forces sent from Europe or the U.S. would be attacked on their way to the region. There was no effective attack by aircraft on our troops or our port and support facilities. Though there were concerns Iraq might employ chemical weapons or biological agents, they were never used."[38]

Overall, observers concluded that Saddam's inability to prevent or even strike at U.S. and coalition forces as they gathered and prepared in his region—and specifically along the Iraq and Kuwait borders and in the waters of the Persian Gulf—doomed his forces to eventual defeat at the hands of strategically superior opponents. Saddam may have barked in his media and propaganda efforts, but he seemed to take no practical action to thwart the grand buildup of high-technology forces and sheer manpower cresting like a wave of steel. In short, he made no effort to conduct a *focused, determined* anti-access campaign that at the very least could inconvenience the buildup. As a study focused on the diplomatic aspects of the Gulf War concludes, "no serious consideration was given to a pre-emptive attack on coalition forces to disrupt and confuse their preparations."[39]

Global Lessons Learned from Desert Storm

Beyond the comments of Soviet and Chinese writers, the consensus that U.S. forces were unbeatable in a conventional force-on-force conflict became the dominant global lesson learned. As the previous study notes, "while the 'Vietnam syndrome' might always have been exaggerated and misinterpreted, the display of U.S. power in the Persian Gulf had the effect of creating an image of overwhelming power."[40] With the concurrent collapse of the Soviet Union, there was no conceivable counterweight military force that could meet the Americans head-on under equal conditions. The PLA had yet to demonstrate an official commitment to developing a modern force, although there were stirrings in that direction.

For nations, nonstate actors, and other entities opposed to U.S. influence and contemplating violent means toward achieving their international political aims, conventional warfare seemed almost a closed option. At the same time, Sundjari's recommendation to develop a nuclear force prior to opposing the United States (or a Western coalition) appeared highly risky. Developing weapons of mass destruction (WMDs) put one in the crosshairs of the world community as a potential member of the axis of evil, not helpful in gaining support or remaining integrated in the global economy.[41] It could be done, and was done by the North Korean regime, but only through "acting crazy" and being willing to forgo all intellectual and economic contact with much of the world—what might be called cognitive anti-access, or anti-access of the mind. Pakistan, another nation that revealed its nuclear capabilities after the Gulf War—following a second Indian nuclear test in 1998—had actually begun its nuclear weapons program in 1972 and avoided as much condemnation as North Korea because it appeared to be balancing a nuclear-armed traditional enemy.

The highly touted revolution in military affairs that was associated with the precision weapon and information systems demonstrated in the Gulf War might conceivably lead to a means of equalization. After all, there was little that the United States and its allies could do about the diffusion of the underlying technical knowledge, a diffusion that would inevitably occur by legal or illegal means. A number of scholars suggested that hostile states could develop *sideways technologies*—in other words, use existing or available military technologies in unexpected ways different from how Western nations were using them. Another perspective was that other nations could develop *disruptive technologies* that could overturn the main advantages of Western forces.[42] But many of the disruptive technologies identified—ballistic missiles, nuclear weapons, cruise missiles, sea mines, space satellites, and drones—had themselves been developed by the Western states, which could likely (if desired) maintain an advantage in technological development and/or eventually develop countermeasures.

If technological development in itself was not a definite answer to enabling success against a U.S. military that had proven its own technological dominance of the battlefield, the question remained: how could U.S. forces be defeated in a conflict? How in a conflict with the United States could potential opponents avoid a result like Operation Desert Storm or the war in Iraq of 2003? For that matter, how could such potential opponents avoid the results of the conflict in Kosovo (expulsion of Serbian forces) or regime change in Libya—conflicts that involved only U.S. naval and air power without extensive ground forces? The apparent answer, based on trends in defense postures, exercises, and procurement, and logical analysis, appears to be the adoption of anti-access principles, the foremost being: do not allow U.S. forces to operate, enter, or remain in the region of conflict. As the National Defense Panel (NDP) of 1997 noted in its independent review of DOD planning, "we can safely assume that future adversaries will have learned from the Gulf War. . . . We can expect those opposed to our interests to confront us . . . with asymmetrical responses to our traditional strengths."[43]

This expectation is explained via logical reasoning in a more recent RAND Corporation report: "The motives for adopting an anti-access strategy are theoretically compelling: If the U.S. military can arrive in force, it will almost undoubtedly win in a conventional military campaign. A rational opponent should thus seek to acquire the capabilities necessary to disrupt or delay U.S. deployment activities or to deny it the use of regional bases in the hope that, by successfully doing so or threatening to do so, it will prevent or deter the United States from acting."[44]

Thus, the most impressive initial victory of modern, high-technology warfare provided not only the justification for potential opponents to pursue modern military technical advancements, but to reconsider adopting a strategic concept utilized at least two thousand years in the past.

Chapter 2

Developing the Modern Concept of Anti-Access

The actual terms anti-access and area denial are decidedly modern. Victory, it is often recalled, has a thousand fathers (or mothers), but defeat always appears an orphan. So it is with modern ideas that turn popular. By 2013 it has become repeated wisdom that anti-access or anti-access/area denial (A2/AD) is the form of conventional warfare the United States would most likely face in a regional conflict. In fact, overcoming anti-access and area-denial challenges was identified as one of the ten "primary missions of the U.S. armed forces" in a January 2012 strategic guidance signed by President Barack Obama himself.[1] The document notes that "to credibly deter potential adversaries and to prevent them from achieving their objectives, the United States must maintain its ability to project power in areas in which our access and freedom to operate are challenged."[2] In support of this strategic rhetoric, the document also provides a directive concerning resource allocation intended to affect future defense budgets: "Accordingly, the U.S. military will invest as required to ensure its ability to operate effectively in anti-access and area denial (A2/AD) environments."[3]

Defining Anti-Access and Area Denial

Previously we have noted that the Department of Defense has established "nearly official" definitions for *anti-access* and *area denial.* These should be considered only that because they have yet to appear in Joint Publication 1–02, the *Department of Defense Dictionary of Military and Associated Terms* (8 November 2010, as amended through 15 August 2011). Also, where the terms appear in the DOD's *Joint Operational Access Concept* (*JOAC*), the authors add the caveat "as used in this paper."[4] The authors also point out that the document is "a warfighting concept," implying that it necessarily avoids much discussion of the political and diplomatic efforts that make the anti-access approach a viable component of grand strategy.

Anti-access is defined in the document as "those actions and capabilities, usually long-range, designed to prevent an opposing force from entering an operational area."[5] The reference to "long-range" is most likely driven by capabilities being developed by such potential anti-access powers as the People's Republic of China (PRC), which include antiship ballistic missiles (ASBMs). But with that exception, this definition correlates with that espoused in this book. Historical examples indicate that "long-range" is an inexact term, and "usually long-range" can be quite elastic. Reflecting the inclusiveness required by current interpretations of *jointness*, the document provides some discussion of the term area denial, taking pains to make a "distinction between antiaccess and area-denial" (the spellings of terms as used in the document).[6] Area denial is subsequently defined as "those actions and capabilities, usually of shorter range, designed not to keep an opposing force out, but to limit its freedom of action within the operational area."[7]

Despite this distinction, the joint concept routinely utilizes the combined term anti-access/area denial, albeit avoiding the A2/AD abbreviation that has become prevalent in commentaries. As noted, this reflects the operational similarities between anti-access and area-denial efforts.

Operational Access

As evident in its title, the *JOAC* is built around the term *operational access*, defined as "the ability to project military force into an operational area with sufficient freedom of action to accomplish the mission."[8] Operational access is conceived as the military contribution to achieving access, particularly in an anti-access environment, or as the *JOAC* describes it, "the joint force contribution to *assured access*."[9]

This is equivalent to the objective of what is referred to in this book as the counter–anti-access force. Assured access is described in turn as "the unhindered national use of the global commons and select sovereign territory, waters, airspace and cyberspace, achieved by projecting all the elements of national power." This comes closer to describing both anti-access and counter–anti-access efforts as elements of grand strategy. Indeed, the *JOAC* provides an excellent justification for addressing the anti-access problem in the broader characteristics of assured access: "As war is the extension of politics by other means, operational access does not exist for its own sake, but rather serves our broader strategic goals, whether to ensure strategic access to commerce, demonstrate U.S. resolve by positioning forces overseas to manage crisis and prevent war, or defeat an enemy in war."[10]

The *JOAC*'s solution to achieving operational access lies in what it terms *cross-domain synergy*, the ability to strike the enemy simultaneously or sequentially from dominant positions in all combat mediums or domains in such a way that operations in each domain provide mutual support for each other. In older tactical terms, this could be described as combined arms warfare. It certainly has the requisite joint flavor. It is indeed a valid method of visualizing the interdependence of the components of both anti-access and countering anti-access, and will be discussed later.

Obviously the *JOAC* sees the problem from the U.S. perspective, the objective of which is the defeat of others' anti-access strategies. This is appropriate, since the terms anti-access and area denial and the understanding of them in a modern context have their roots in a strategic problem faced by the United States during the Cold War.

Naval Roots of Anti-Access

Our modern understanding of naval strategy dates from the turn of the twentieth century, when the great theorists of sea power such as Capt. Alfred Thayer Mahan and Sir Julian Corbett argued over the purposes of navies. Part of the debate stemmed from the fact that the United States appeared to have the potential to supplant Britain's Royal Navy as the dominant naval power and the guarantor of international freedom of the seas.

The debates involved many ideas, some subtle and appreciated only by naval professionals, but they all focused on the concept of *command of the seas* and its importance or relative unimportance. Mahan, who also was also the first to describe the sea as a global common, emphasized that a dominant navy that could command the sea had the power not only to defeat other navies, but to control the flow of international trade. This was not an idea that he originated, and his name is also associated—in a somewhat exaggerated fashion—with the need for decisive battle in a naval war to achieve true command. Centuries before, Sir Walter Raleigh wrote that "whosoever commands the sea commands the trade; whosoever commands the trade of the world commands the riches of the world, and consequently the world itself."[11]

But Mahan was the great articulator of sea power and was able to explain the role of navies in terms of denying access to the global common on which the trade of nations (or trade beyond their immediate neighbors) relied. This was a geoeconomic purpose as well as military one, and when articulated today the term *globalization* inevitably becomes involved. Rhetoric of the 1890s was a bit more flowery than today, but perhaps his most cogent observation was: "It is not the taking of individual ships or convoys, be they few or many, that strikes down the money power of a nation; it is the possession of that overbearing power on the sea which drives the enemy's flag from it, or allows it to appear only as the fugitive; and by controlling the great common, closes the highway by which commerce moves to and from the enemy's shores." To this he added: "This overbearing power can only be exercised by great navies."[12]

Although Mahan described it from the perspective of trade, he also emphasized that access to the sea and access to the land from the sea were the hinge upon which great conflicts of the past turned. Returning to our example of the Greek-Persian War, one sees that, when cut off, access to the sea denied Xerxes the ability to maintain his land forces. He retreated from his out-of-area operation. In addition, access to the land from the sea was also access to the enemy's coastline and surrounding regions, an access that could be exploited to directly influence events on land—particularly by overwhelming navies with power-projection capabilities. The idea that navies could have significant effects ashore became institutionalized after Mahan, particularly during the development of specialized amphibious warfare. But from his time naval warfare became associated with a struggle for access.

Other theorists saw the struggle for sea power more in terms of a straightforward military problem: how to defeat an enemy's fleet or how to prevent being defeated by a stronger fleet. In this they would have probably appreciated the current *JOAC*, with its deliberate focus as "a warfighting concept" and emphasis on operational access.[13] Influenced by their perspective and later naval thinkers, the term *command of the sea* began to be replaced by the less grandiose term of *sea control.*[14] Sea control was what great navies sought in war: the ability to prevail in the area of conflict and utilize that sea-space (and airspace) for its own purpose. Conversely, *sea denial* was what the lesser navies could hope to achieve: deny their opponent the free use of that sea-space for its purposes—the creation of a naval no-man's-land (or, rather, no-man's-sea). If our earlier nearly official definition of the joint term *area denial* seems quite similar, it is because it is almost undeniably derived from *sea denial*, a term used in military literature for at least fifty years before.[15]

The Cold War Sea-Denial and Anti-Access Problem

For much of the Cold War, the Soviet navy could not compete with the U.S. Navy and allied navies for sea control. The United States ended World War II with thousands of ships and an experience with intensive naval warfighting that could not be duplicated. Two particular areas in which the United States had an insurmountable advantage over all other navies were aircraft carrier air power and amphibious operations. In contrast, naval operations had been peripheral for the Soviet Union in World War II, and as a traditionally continental power its Russian predecessor did not have a distinguished naval tradition.

Faced with U.S. and allied navies that could control (or command) the oceans of the world, and particularly the Atlantic, the Soviet navy was developed as a sea-denial force—one that would attempt to destroy enemy ships and aircraft but did not intend to control sea regions far beyond its immediate sea frontiers, which were the Barents Sea, the White Sea, the Black Sea, and the Sea of Okhotsk. Instead of (initially) building aircraft carriers or large surface combatants, the Soviet navy invested in a portfolio of what even today would be considered the tools of anti-access: submarines, both diesel and nuclear; long-range bombers; air-, ship-, and submarine-launched cruise missiles; sophisticated naval mines; and nuclear ballistic missiles targeted, at least in rhetoric, at ships. As noted, later in Admiral Gorshkov's tenure the Soviet navy began to construct ships more suitable to the sea-control role, such as the *Kiev*-class vertical and/or short take-off and landing (V/STOL) aircraft carrier and *Kirov*-class nuclear-powered cruiser. But for much of its existence, the Soviet navy was focused on sea denial.

The Soviet sea-denial role was directed at the attrition of U.S. carrier battle groups and the destruction of the transatlantic resupply effort if a war with NATO forces on the central front in Germany were to occur—an anti-access effort determined to sever the prodigious armaments and logistical capability of North America from NATO Europe. The Soviet sea-denial threat was indeed a formidable one; even nuclear-armed cruise missiles carried by long-range bombers were part of this force structure. For much of the 1960s and 1970s, the U.S. Navy and allied navies—along with elements of the U.S. Air Force—focused on a maritime anti-access effort of their own: preventing Soviet submarines from crossing the sea region between Greenland, Iceland, and the United Kingdom known as the GIUK gap (or barrier) and reaching interdiction positions in the open Atlantic. But development in naval technologies in the 1980s and the Reagan administration's efforts to build a six-hundred-ship navy inspired a different approach—the famed (or notorious to its critics) Maritime Strategy in which the U.S. Navy and Marine Corps—with support from U.S. Air Force units not part of the primary AirLand Battle effort in the central front—would seek what today would be called operational access into Soviet home waters and adjacent coastline.

The Maritime Strategy was publicly released in 1986.[16] Critics of the Maritime Strategy portrayed it as likely to cause a "conventional" Soviet-NATO war to go nuclear because it would threaten the sea "bastions" in which Soviet nuclear ballistic missile submarines—supposedly the Soviet Union's primary strategic nuclear deterrent—were concentrated. In this they missed an unspoken assumption that many in the U.S. Department of Defense, and particularly in the U.S. Navy, were beginning to believe: NATO would likely lose the force-on-force battle in the central front against the ever-strengthening Warsaw Pact forces. In its warfighting strategy, NATO would be forced into the position of itself using tactical nuclear weapons or conceding Central Europe, possibly up to the Rhine. No one knew what the Soviet reaction would be to the use of tactical nuclear weapons. But in an attempt to avoid this scenario and force the Soviet Union to take Warsaw Pact forces away from the central front in order to protect their periphery, U.S. naval leaders were willing to risk carrier battle groups in the Northern European fiords and in the teeth of Soviet naval anti-access defenses.

From this perspective it can be seen that the NATO naval campaign of the Cold War was largely an access versus anti-access struggle as the Soviets sought the capability to sever the United States and Canada from reinforcing the European region, and NATO sought to gain access to the Soviet periphery. This was a vastly different type of campaign than what the U.S. Army and U.S. Air Force faced in central Europe, a potential conflict across a static front, with fixed land bases to supply land and air forces into largely an attrition battle. To Army and most Air Force planners, as well as most secretaries of defense and the majority of think tank pundits, the naval struggle for operational access to the periphery of the Soviet Union was quite peripheral. In contrast, U.S. Navy and Marine Corps operational planning envisioned a classic struggle to break an opponent's anti-access strategy, even if it was not articulated as such. The fact that the Soviet Union designed its naval forces for a sea-denial and anti-access mission is a key to understanding the development of anti-access as an independent concept.

Enter the Office of Net Assessment

Among students of U.S. military strategy, the Pentagon's Office of Net Assessment has earned legendary status. Run for more than four decades and during both Republican and Democratic administrations by Andrew Marshall, a long-range thinker who has never sought public attention, Net Assessment was indeed established to conduct net assessments—analyses of the relative power of the United States versus potential opponents, particularly the Soviet Union during the Cold War.[17] Much of its work was and is classified, and few of its actual papers reach the public, but many of the concepts formulated within were subsequently extolled in public form by its civilian and retired military alumni. Even today Marshall has the knack of picking some of the broadest thinkers within the DOD for his staff and commissioning studies by rising defense intellectuals from universities and think tanks. From them, original and unconventional ideas about defense naturally flow, even when the analysts return to teaching and think-tanking.

Sometime in the late 1980s, Marshall turned the attention of part of his staff to analysis of the publicly controversial Maritime Strategy. His essential question was whether the U.S. Navy and Marine Corps and other NATO forces could actually carry out a successful naval campaign against the Soviet periphery. Critics argued that most of the U.S. ships would be sunk, particularly aircraft carriers—often portrayed as sitting ducks—and Marshall wanted to study the problem away from the Navy's self-justifying public arguments and the editorial battles of the critics. This study, including sponsored war games, was but one of many that Net Assessment was doing simultaneously, others focusing on Soviet industrial and military capabilities for global war and on the potential for a military-technical revolution (MTR), later known as the revolution in military affairs. It was not by any means considered the most important study then being carried out.

The Soviet Union may have fallen in the early 1990s, but the question of whether the U.S. Navy could operate close to a well-armed opponent remained. Results of related Net Assessment studies and war games were briefed to the Chief of Naval Operations staff—known as OPNAV—and other organizations in the DOD. A notable briefer was Capt. Patrick Curry, USN, who had been the head of the OPNAV Strategic Concepts Branch (later renamed the Strategy and Concepts Branch) before being assigned to the Office of Net Assessment. The Strategic Concepts Branch had gained fame for its role in drafting *The Maritime Strategy*. To say that Net Assessment studies, some of them skeptical of naval capabilities, were received with a certain amount of ill favor by the Navy leadership would be an understatement. Whatever its formal name, the Net Assessment–originated study became known as the "Anti-Navy Study." One could argue that such a title was innocent enough, since it was concerned with Soviet capabilities versus the U.S. Navy. But in the Pentagon atmosphere of struggles for defense resources, "anti-Navy" was certainly provocative wording.[18] Somewhere along the voyage, it was conceded that because the study was about the Navy's capability to achieve access to the Soviet periphery, "anti-access" was a more suitable description. As far as can be determined, this was the first actual use of the term.

Popularization of the Term

An independent concept of anti-access started to gain traction among naval thinkers about the same time as Operation Desert Shield/Desert Storm, the Warsaw Pact's dissolution, and the Soviet implosion. These were rather profound events. The Maritime Strategy was obviously overtaken, and the Navy shifted to articulating its ability to provide for expeditionary warfare and access for potential intervention in the many regional crises that routinely occur. This . . . *From the Sea* approach largely put the U.S. Marine Corps on equal status with the bluewater Navy or at least closer to it than the roles envisioned in the Maritime Strategy. After all, it would be the Marine Corps, with its long tradition of being America's force of choice for "small wars," that would be doing the intervening in the conflicts, with the Navy's primary role being that of getting them there and providing air support, fire from the sea, protection from maritime threats, and logistics. Since this role could be summed up as providing access, the terms began to be used more and more by the Navy and Marine Corps in officially describing their contribution to the joint force, jointness having been enshrined in the DOD by the organizational changes in response to the Goldwater-Nichols Act of 1986. But the concept of access remained largely confined to a naval audience, only reluctantly being admitted to joint discussions.

However, these discussions slowly began to include concerns about the viability of American access to the trouble spots of the world, most of which were within range of power projection by naval forces. Perhaps the first obvious threat came from naval mines, which played a significant role in dissuading Gen. Norman Schwarzkopf from authorizing a Marine landing on the coasts of Kuwait and Iraq during Operation Desert Storm. Mine countermeasures had always been a neglected mission in the U.S. Navy because of its apparent pedestrian nature and limited impact on open-ocean operations. Even under the . . . *From the Sea* strategic vision, resources assigned to mine countermeasures were minor. Likewise, interest in and resources for naval mining by the United States were practically nonexistent, relying as it did on an aging stock of mines to be delivered largely by U.S. Air Force B-52s. But with the shift to a littoral focus that emphasized the Navy and Marine Corps expeditionary nature, threats to such expeditions needed to be acknowledged in the joint arena. Still, the overall focus on threats to operational access remained decidedly naval rather than joint: naval mining; submarines, particularly in narrow straits; air-, sea-, and land-launched cruise missiles; fast patrol craft; long-range bombers; and ballistic missiles, perhaps carrying WMDs.[19] For opponents with more limited resources, threats could include swarms of small craft with shorter-range, possibly handheld missiles and rockets; ground combat weapons adapted for naval purposes; less-sophisticated or improvised naval mines; combat swimmers; and suicidal aircraft (discussed prior to 9/11 and reflecting the legacy of the kamikaze).

In a way, the focus on naval access was a return to Mahan, with his emphasis on decisive battle removed. Within discussions of globalization, the role of the Navy in ensuring the economic access of the United States to markets and sources of raw materials was also being acknowledged.[20]

Initial Joint Reluctance and the RMA

The Department of the Navy's focus on anti-access threats and its persistent public reference to its expeditionary nature and ability to provide presence in regions where U.S. forces did not have access to land bases was initially greeted with defensiveness by the Army and Air Force. Having relied on forward land bases overseas throughout their positioning in the Cold War, neither service had previously advertised itself as an expeditionary force. Moreover, the potential threats being discussed as tools of an opponent's anti-access strategy—notably ballistic and cruise missiles—would be even more effective against fixed land bases of known coordinates than they would against targets maneuvering at sea. Moving targets required sensors and battle management systems for locating and solving a changing fire-control problem. Immovable land bases did not, and their vulnerability against any enemy that could reach them was painfully obvious. Efforts could be and were made to defend land bases against attack, but saturation missile attacks, particularly bearing WMDs, made long-term defense appear problematic.

However, since there seemed to be no potential enemy with such capabilities in the immediate post–Cold War, post–Desert Storm "new world order," the question of access of U.S. forces to such areas as Northeast Asia, Southwest Asia, and Europe could be dismissed as a bit of naval parochialism. As Maj. Gen. Charles D. Link, USAF, special assistant to the Air Force chief of staff, stated on 3 January 1995, in a letter to the congressionally mandated Commission on Roles and Missions of the Armed Forces, "with regard to [Chief of Naval Operations] Admiral Boorda's concern about [USAF statements concerning] 'unlimited access to foreign basing or that an enemy will not attack the airfield we intend to use' we are frankly perplexed. Since the establishment of the United States, we know of no significant operation in which landbased airpower has failed to contribute because of basing constraints."[21] Later in the letter, General Link continued to say that "we would disagree with the implications that airfields are somehow more vulnerable to attack."[22]

However, such dismissal bumped against another emerging defense concept—also originating in the Office of Net Assessment—that a revolution in military affairs was occurring in which the high-technology type of weapons demonstrated by U.S. forces in Desert Storm would eventually be obtained by potential opponents as the underlying technologies became more diffuse. This was another long-term concept studied by Net Assessment that had been derived from 1980 Soviet sources postulating that a military-technical revolution was being brought about through the use of computers and other advances in command and control, communications, intelligence, surveillance, and reconnaissance—commonly referred to be the abbreviation C^4ISR. Precision weapons were the result of such advancements. By the 1990s Marshall was convinced that this was a valid conceptualization of the changes in warfare brought about by precision, and the American term was soon to be publicized by his alumni following Desert Storm.[23] Although Desert Storm was deemed a great success, many argued that U.S. armed forces needed to be *transformed* in order to capitalize on the new capabilities, keep technologically ahead of potential opponents (who would likely not be as inept as Saddam), and reduce American vulnerability resulting from the diffusion of modern military technologies. Through congressional commission and think tank studies, publicity concerning the RMA and the argument for defense transformation soared. It would become a centerpiece of defense planning during the early 2000s under Secretary of Defense Donald Rumsfeld. But in an interesting fashion, belief in an ongoing RMA also made a shift toward anti-access strategies by potential opponents even more logical.

If indeed potential opponents could develop high-technology weapons (or even not-quite-as-high-technology weapons) through what was earlier referred to as sidewise technologies, the reach and accuracy against fixed targets such as airbases would certainly increase. This issue is also one concerning international politics and diplomacy. Most U.S. forward land bases are located in the territory of allies or partner nations, and their existence and use is ultimately dependent on that nation's permission. Speculation began to circulate that other nations might be reluctant to allow U.S. forces to use their territory if it subjected them to direct attack by a powerful opponent. In contrast, naval forces afloat remain under the complete sovereignty of their flag while operating on the high seas in accordance to international law, requiring no other nation's permission. This was and is a serious issue, since there are examples of the use of U.S. forward bases being restricted by the host nation, notably by Turkey in Desert Storm. But reaction to this suggestion was initially steeped in some parochialism. In response to a conference question, Gen. John P. Jumper, USAF, then commander of U.S. Air Forces in Europe, maintained that "access is an issue until you begin to involve the vital interests of the nation that you want and need as a host. Then access is rarely an issue."[24]

Congressionally Mandated DOD Studies and Defense Transformation

Undoubtedly a factor in propelling the anti-access concept—if not the term—into serious joint discussions was the series of DOD studies in the later 1990s, some of them congressionally mandated, intending to define the future security environment. One was a study on strategic mobility by the Defense Science Board (DSB) in 1996 that focused on the premise that "future adversaries will have the motives and likely the means to seriously disrupt U.S. strategic deployments."[25] As interpreted by the RAND Corporation, a federally funded research and development center (FFRDC) based in Santa Monica, California, the DSB's

1996 report was significant in emphasizing such means of disruption as "coercing other states into not cooperating with U.S. deployment efforts or by directly attacking ports and airfields; [and attacking the following:] logistical nodes; strategic transport assets; and command, control, communications, and computer systems." The means would include "missiles, mines, special operations forces (SOF), aircraft, submarines, offensive information warfare, weapons of mass destruction (WMD), or advanced conventional weapons."[26] RAND's conclusion was that "this early characterization of the anti-access problem is broadly reflected in most subsequent analyses of the issue," which is indeed a fair statement. The DSB's 1996 report also reflects the start of a serious DOD interest in the anti-access problem and its movement toward seeing it as more than just a naval problem.

Although the DOD's *Quadrennial Defense Review* (*QDR*) of 1997, a congressionally mandated report, did not directly refer to the anti-access concept, it did discuss some of the techniques identified by the DSB under its discussion of *asymmetrical threats*, another concept that had gained intellectual (and official) traction.[27] However, members of the Congressional Armed Service Committees were skeptical that the DOD could perform an extensive review of emerging defense threats without defending its existing programs and required the secretary of defense to charter an independent review of the *QDR* by a panel of outside experts (or at least outside of the DOD at that time). This review group was the previously mentioned National Defense Panel. The report it produced—the drafting of which was supported by a staff drawn from the DOD—was titled *Transforming Defense: National Security in the 21st Century*, certainly giving an obvious clue to its perspective and conclusions.

Like the 1997 *QDR*, the NDP report does not use the term anti-access but does refer to "access" in describing capabilities of, and future threats to, America's *power projection* capabilities—the ability to project military power into far regions in order to affect crises and other events. The NDP report states the future problem rather succinctly:

> The cornerstone of America's continued military preeminence is our ability to project combat power rapidly and virtually unimpeded to widespread areas of the globe. Much of our power projection capability depends on sustained access to regions of concern. Any number of circumstances might compromise our forward presence (both

> bases and forward operating forces) and therefore diminish our ability to apply military power, reducing our military and political influence in key regions of the world. For political (domestic or regional) reasons, allies might be coerced not to grant the United States access to their sovereign territory. Hostile forces might threaten punitive strikes (perhaps using weapons of mass destruction) against nations considering an alliance with the United States. Thus, the fostering and nurturing of allies and alliances, as well as our ability to protect our allies from such threats, will be an important factor in our future ability to project combat power anywhere in the world.[28]

The NDP also acknowledged the vulnerability of land bases and reliance on host nation infrastructure as a threat to access: "Even if we retain the necessary bases and port infrastructure to support forward deployed forces, they will be vulnerable to strikes that could reduce or neutralize their utility. Precision strikes, weapons of mass destruction, and cruise and ballistic missiles all present threats to our forward presence, particularly as stand-off ranges increase. So, too, do they threaten access to strategic geographic areas. Widely available national and commercial space-based systems providing imagery, communication, and position location will greatly multiply the vulnerability of fixed and, perhaps, mobile forces as well."[29]

The NDP report had a significant impact on defense debates. It helped "legitimize" and publicized the concepts of the RMA and the need for defense transformation. It also opened the door to commentary and other studies that began to use the term anti-access to describe potential threats to U.S. power projections into regions of concern. The widespread use of the term in official U.S. Navy documents and independent professional journals such as U.S. Naval Institute *Proceedings* led it to being considered for inclusion in overall joint service doctrine.[30] By the late 1990s, the Air Force also decided to brand itself as an expeditionary force, reorganizing into expeditionary air wings and explaining its missions in terms of maintaining overseas access. At the same time, it began to encourage studies within its senior educational institutions, such as the Air War College, and its research contractors, notably the RAND Corporation, to seriously examine the base vulnerability issue.

One particular NDP member, retired Army colonel Andrew F. Krepinevich Jr., also an Office of Net Assessment alumnus, later popularized "A2/AD."

Enter the Think Tanks

One cannot underestimate the influence that the many documents generated by private and public policy research institutes—the think tanks—have on the American defense debate. These documents can shape official American defense in direct and indirect ways. The most direct occur when think tank scholars are appointed as civilian officials in the DOD and bring their ideas with them to implement or at least attempt to implement. The indirect way is in influencing public discussions effectively enough to encourage the DOD to adopt their language and concepts. The political spectrum of think tanks, many officially neutral but still retaining some political biases, ensures that both supporters and critics of DOD policies have intellectual ammunition in which to propel their views with Congress or the administration. Usually included under the category of think tank are the FFRDCs that do research on direct contract with the U.S. government.

It is hard to sort out which think tank was the first to use the term anti-access in describing the challenges to U.S. military power projection. But one that utilized it extensively from about 1999 onward was the Institute for National Strategic Studies at the National Defense University—most obviously an in-house DOD center, but one filled with scholars and military officers ostensibly enjoying a degree of academic freedom. One particular project, a preparatory study intended to provide intellectual capital for the forthcoming *QDR* of 2001, generated several publications tying anti-access strategies to asymmetrical warfare and identifying it as a standard feature of the future security environment.[31] Another project, investigating the effects of globalization—the hottest topic in foreign policy at that time—on U.S. defense policies and posture, resulted in an edited volume bringing the anti-access debate back to its naval roots but with contending opinions on whether potential opponents were investing in specific anti-access weapon systems.[32]

RAND, the Persian Gulf, and China

The RAND Corporation has played a continuous role as strategic consultant to the U.S. government since the 1950s. Along with consulting with a number of federal agencies, RAND has direct, long-term contracts with both the Air Force and Army (as separate programs) to provide research and analysis in operations, personnel policies, force structure, and service strategy. Its two major publicly available reports referencing anti-access strategies were both undertaken by RAND's Project Air Force.

The first, published in 2002 and completed before the invasion of Iraq, is an assessment of the ability of the U.S. Air Force to carry out an early halt mission—quickly interdicting and defeating a mechanized invasion by Iraqi forces in Saudi Arabia or elsewhere. Titled *Measuring Interdiction Capabilities in the Presence of Anti-Access Strategies: Exploratory Analysis to Inform Adaptive Strategy for the Persian Gulf*, it provides a typology of potential actions and counteractions related to anti-access strategies, which it defines as strategies "that would impede American use of regional bases and otherwise hinder efficient operations in the region."[33]

The study particularly emphasizes the threat of WMDs and deception as key elements of anti-access, but it is also one of the few that overtly acknowledges the impact of geographic factors, probably because its analysis is specifically focused on a desert environment. It notes that "these elements [enemy scheme of maneuver, WMDs, deception] of an anti-access strategy are under the control [of the anti-access force], while others are more intrinsically related to the particular theater's terrain or to chance."[34]

RAND's second major anti-access study, published in 2007, tackles what most defense analysts view as the most formidable anti-access effort, that of the People's Republic of China. Titled *Entering the Dragon's Lair: Chinese Antiaccess Strategies and Their Implications for the United States*, the study examines broad categories of anti-access techniques and weaponry, as well as details specific actions the People's Liberation Army could take. The study is sourced from professional writings by PLA officers and associated academics, making an effective case that the PLA currently views potential hostilities with the United States through an anti-access lens. In its own documents, the PLA calls its anti-access strategy *Shashoujian*, or "Assassin's Mace," a folkloric reference to a club used by a hero, notably playing by his own rules, to overcome a more powerful enemy.[35] The RAND authors maintain that "Chinese strategists appear to understand the success against the U.S. military depends on China's ability to avoid a direct confrontation with U.S. forces in a traditional force-on-force battle."[36]

Buttressing the attractiveness of the anti-access approach is a view that factors included within what has earlier been described as the fundamental elements of anti-access could mitigate an advantage in high-technology weaponry. In the words of the study, "underlying the Chinese approach toward a potential conflict with the United States is the conviction that even an adversary with superior weapons, technology, and equipment will be unable to maintain absolute superiority in all respects. Military conflict does not occur in a vacuum. Rather, it takes place within a specific geographic and political context that will inevitably provide the PLA with opportunities to offset the United States' advantage in technology."[37]

If one adds this conviction to a perspective that the PRC is a "rising power" whereas America is "in decline," there could be a tendency by PLA planners to assume an even greater degree of the effectiveness of their anti-access tools than they might warrant if a hostile confrontation occurred. This rising power/declining power perspective has caused considerable concern among the PRC's neighbors who see the United States as providing balance in the region.[38]

It is reasonable to assume that RAND has also conducted considerable classified analysis for the DOD on anti-access issues.

CSBA and A2/AD

Even more effective in popularizing the anti-access concept has been the work of the Center for Strategic and Budgetary Assessments (CSBA), an institute that combines the features of think tank and defense consultancy. CSBA has been headed since the 1990s by Andrew J. Krepinevich Jr., a prominent Office of Net Assessment alumnus and, as previously noted, a NDP member. CSBA's papers and briefings have influenced Democratic and Republican administrations alike, and it is the research center most associated with the concept, having prepared the most public studies concerning the anti-access threat to U.S., NATO, and other allied interests.

With a long career as a soldier-scholar, Krepinevich has a public reputation that began with the publication in 1988 of his book *The Army and Vietnam*, in which he argues that senior Army and civilian leaders never understood what kind of war they were fighting in Southeast Asia and that no amount of resources could have made their approach successful. The book caught attention because of the sources used—not the op-ed opinions of critics, but the Army's own declassified documents and interviews with combatants. A contemporary of Captain Curry, Colonel Krepinevich served in Net Assessment in the early 1990s where he was a lead analyst of the RMA and one who brought that and the term defense transformation into public prominence. He would do likewise to the term anti-access/area denial, which he shortened for convenience to A2/AD.

CSBA issued its first public papers on anti-access strategies in 2002. One, on the vulnerability of overseas air bases in an anti-access environment, was contracted from Christopher J. Bowie, then a senior analyst at the Northrop Grumman Analysis Center and former RAND analyst and member of the secretary of the Air Force's staff. From his position as a trusted agent, Bowie is able to point to both the operational and political vulnerability of forward land air bases in a way acknowledged by U.S. Air Force leaders. On the political vulnerability, his concern is uncertainty: "The attitude of host countries regarding access in future crises is difficult to predict, raising significant uncertainties regarding the basing and employment of combat aircraft. The United States can bring enormous pressure to bear on a host country to accept US forces, but success, as has been seen in numerous crises, cannot be guaranteed."[39]

Adding the operational anti-access challenges, he makes the case for greater Air Force investment in longer-range aircraft, arguing that recent budgets were skewed more toward short-range fighters than in previous Air Force budget planning. He concludes: "This report suggests that over the long run, the combined uncertainties raised by political factors, logistics, and emerging military threats mean that the combat power of the land-based fighter force may be significantly constrained in supporting US power-projection operations in an anti-access environment. To hedge, the Defense Department should adjust its current combat aircraft modernization plans, which focus primarily on the acquisition of fighter aircraft, to increase spending on systems less reliant upon forward bases."[40]

In a subtle push against Department of the Navy claims that naval forces can conduct extensive forward operations without dependence on overseas bases, Bowie introduces the base vulnerability into a joint context: "The issues raised in this analysis have broader strategic implications for the US military as a whole. Reliance on large, fixed facilities in the theater of operations is much more than an Air Force issue."[41]

The second CSBA report issued in 2002 defends, in part, the Department of the Navy's claim to operational independence of its forward-deployed forces from basing, while arguing that the Navy and Marine Corps need to become full participants in the defense transformation. Written by retired Marine colonel and subsequent undersecretary of the Navy Robert O. Work, *The Challenge of Maritime Transformation: Is Bigger Better?* is an overall assessment of then current Navy–Marine Corps force structure rather than a study of anti-access per se.[42] But Work, perhaps influenced by the Navy's description of its decentralized mode of operations as network-centric warfare, introduces the idea of *anti-access networks* to describe a potential adversary's overall system of sensors, electronic measures and countermeasures, and offensive and defensive weaponry that could be used to support an anti-access strategy.

Anti-access network is obviously Work's preferred method of describing the anti-access challenge to U.S. forces; he uses the term 93 times in his 169-page report. It is indeed an effective way to shift away from possible fixations on particular anti-access weapon systems—PRC antiship ballistic missiles, for example—toward an understanding that if an anti-access strategy is to be effective, an entire interdependent network of systems and actions are required to work together successfully. But this also means that anti-access strategies are vulnerable to disruption if the elements of the network become disaggregated. Countering anti-access strategies thereby consists of attacks against an organic whole in order to break apart the components so that they can be destroyed in detail or bypassed—in other words, to break the wall.

In 2003 CSBA released a coauthored report assessing individual service plans and programs in terms of their effectiveness in an anti-access and area-denial environment. Titled *Meeting the Anti-Access and Area Denial Challenge*, it was here that the acronym A2/AD was first used extensively. Part of the logic behind this lay the inability to clearly and effectively distinguish between anti-access and area denial, despite whatever nuanced definitions could be used. The tactics and weapon systems appropriate to countering anti-access strategies appear to be the same appropriate for countering area-denial defenses. Actions taken by the defender seem to vary only in terms of the ranges of the weapons involved. But another significant part of the logic was the desire to be able to assess Army programs against anti-access–type systems. Since America's geographic position clearly mandates the predominance of the maritime domain in ensuring out-of-area access in the face of extensive anti-access networks, it is difficult to postulate a role for the Army in the counter–anti-access mission. Krepinevich and his colleagues make substantive efforts to utilize a joint perspective—as currently understood—in their analyses. Using the compiled A2/AD to do this is capable of being inclusive without generating debate as to which types of forces are most suitable for countering an opponent's efforts to keep U.S. forces away from its region and without, presumably, an even greater struggle for DOD resources between the services.

Meeting the Anti-Access and Area Denial Challenge, in which Barry Watts assesses the Air Force's Global Strike Task Force vision and its reliance on the F/A-22, Robert Work appraises the Navy–Marine Corps claim to providing assured access and the decision to base future shipbuilding on the Littoral Combat Ship (LCS), and Andrew Krepinevich critiques the Army's Objective Force and the development of lighter combat units, does make some excellent observations concerning anti-access and the challenge it presents to existing U.S. force structure.

The report makes the point that the 2003 crushing of conventional Iraqi forces reinforced the incentive for future opponents to adopt an anti-access approach and returns to the NDP observation that anti-access tactics can be described as asymmetrical warfare: "There is ample reason to anticipate that future adversaries, having seen Iraq routed twice by US-led coalition forces after they were allowed to deploy unmolested into Southwest Asia, will seek asymmetric ways of opposing the movement of US military forces into their region."[43]

The report also emphasizes the proliferation in high-technology sensors—the targeting systems so essential in directing the high-technology weapons themselves and operationalizing the fundamental element of the criticality of information. The report states: "Disconcerting is the growing proliferation of national and commercial satellite services and missile technology. Increased access to these satellite services will allow even regional rogue states both to pretarget key fixed facilities and to monitor US deployments into forward bases. Unless one makes heroic assumptions regarding advances in missile defense effectiveness—which this assessment does not—these facilities can be held at risk through the employment of even moderate numbers of ballistic and cruise missiles. This is particularly true if an adversary has and threatens to use missiles with chemical, biological, radiological, nuclear, or enhanced explosives warheads."[44]

The report also bemoans the fact that the individual services were pursuing their own individual force structure changes (or perhaps new force structure depictions) to deal with their perceived slice of anti-access challenges. This is a major theme of subsequent CSBA reports and the basis for their advocacy of the AirSea Battle construct. In its own words, "the disconnects between individual Service solutions to the A2/AD challenge, then, are substantial. Furthermore, these disconnects suggest an obvious recommendation. A joint approach to the prospective A2 and AD capabilities of future US adversaries is crucial if the various path, operational, technological, and fiscal risks are to be mitigated or hedged against to any serious degree."[45]

However, the report ultimately seems a paean to the reigning imperative for defense transformation rather than the assessment of optimal programs for counter–anti-access. The service force structure programs assessed and the manner of their assessment hinge on the degree to which they constitute transformation and could likely be examined in a similar fashion without reference to anti-access.

AirSea Battle and Its Interpretations

In September 2009 the Air Force chief of staff, Gen. Norton A. Schwartz, and the then chief of naval operations, Adm. Gary Roughead, signed a classified memorandum to begin biservice developing of a new operational concept known as "AirSea Battle." Officially this agreement was directed by the then secretary of defense, Robert Gates, but whether this initiative was initiated by the services themselves, by the Office of Secretary of Defense (OSD) staff, or through the advice or consultation of CSBA or another outside group is unclear. The memorandum led to the establishment of an interservice AirSea Battle Office. Initially, very little was publicly revealed concerning the deliberations or anticipated product (in DOD terms, "the deliverable") to be generated by the office, but it was assumed by outside analysts that extensive work was being done on the classified level, to be made public several years into the effort. This would follow the model of the Maritime Strategy. The premise, however, was clear: to create a coordinated approach by both services in aligning doctrine and—possibly—force structure to deal with the emerging anti-access threat from the PRC, Iran, and North Korea. According to the *JOAC*, the "Air-Sea Battle is a limited operational concept that focuses on the development of integrated air and naval forces in the context of anti-access/area-denial threats. The concept identifies the actions needed to defeat those threats and the materiel and nonmateriel investments required to execute those actions."[46]

In describing the expected result of the AirSea Battle Office effort, the *JOAC* refers to three components or commitments:

> There are three key components of Air-Sea Battle designed to enhance cooperation within the Department of the Air Force and the Department of the Navy. The first component is an *institutional* commitment to developing an enduring organizational model that ensures formal collaboration to address the antiaccess/area-denial challenge over time. The second component is *conceptual* alignment to ensure that capabilities are integrated properly between Services. The final component is doctrinal, organizational, training, materiel, leadership and education, personnel, and facilities *initiatives* developed jointly to ensure they are complementary where appropriate, redundant when mandated by capacity requirements, fully interoperable, and fielded with integrated acquisition strategies that seek efficiencies where they can be achieved.[47]

The last statement concerning efficiencies sent a bit of a chill to analysts who view anti-access as a particularly significant and difficult threat requiring a resource-intensive response. At the announcement of the initiative, some suspected that creation of the biservice effort masked an OSD-directed ambition to cut or reduce Navy or Air Force acquisition programs under the justification that they were redundant in dealing with the anti-access threat. The facts that the initiative was identified as "directed" by Secretary Gates and that he was trying to find $100 billion in defense savings over the next five years in order to instill a "culture of savings and restraint" in the military intensified the supposition that the effort was really—in Pentagon vernacular—a downsizing "budget drill." This has never been formally denied by OSD despite the fact that General Schwartz and follow-on CNO Adm. Jonathan W. Greenert have stressed that "with Air-Sea Battle, we are reinvigorating the historic partnership between our two departments to protect the freedom of the commons and ensure operational access for the Joint Force."[48]

But while the official objective of the AirSea Battle effort may be clear in rhetoric, the intended result remains murky. As one critical review remarks, "in 2009, the Navy and the Air Force introduced a new concept for fighting called 'Air-Sea Battle.' Three years later, even Pentagon insiders are still completely baffled by it."[49] This puzzlement appears to stem from anticipation that the proximate product would be a white paper or doctrinal statement on how the services would jointly conduct a counter–anti-access campaign, particularly against the toughest challenge—the People's Liberation Army (PLA)—even if China, for political sensitivities, could not be referred to by name. It is uncertain whether such a statement was ever written, whether in general or detailed. At a press brief in November 2011, a Pentagon spokesman stated that "the secretary of defense has acknowledged the work as credible work and has given us the green light to move forward with the implementation of the Air-Sea Battle concept."[50] Perhaps the simple result is the inclusion of anti-access as one of the "Primary Missions of the U.S. Armed Forces" in the January 2012 strategic guidance. If a dedicated document exists, the sensitivity acknowledged previously may preclude public release. In fact, former vice chairman of the Joint Chiefs of Staff Gen. James Cartwright, USMC, stated rather critically in an April 2012 conference that "AirSea Battle is demonizing China. That's not in anybody's interest."[51]

It should be acknowledged that a biservice plan must necessarily be focused on doctrine and force structure, since the authority and responsibility to develop an actual campaign or war plan is legislatively and administratively vested in the regional combatant commander, which in the case of the PRC would be the head of the U.S. Pacific Command. In the case of Iran, it would be the head of the U.S. Central Command. Under congressional reforms, the service chiefs are not in the "strategy business" except in terms of the strategic vision of the services themselves—a vision of what the forces that they are responsible for building, training, and providing to the combatant commanders are designed to do. In this, it would be perfectly legitimate for the services to identify the systems, training, and doctrine that exist or are in development to deal with the general anti-access threats that the combatant commanders might face. Based on the forces anticipated to be assigned to the theater, the combatant commander can draft the war plan. Service visions need not specify a particular potential opponent, even if their strategies or capabilities are discussed in general terms.

It can be argued that the *JOAC* itself can function as the biservice (as well as the joint) vision. In fact, Air Force and Navy planners were involved in its drafting, along with the Army and Marine Corps and the Joint Staff. But the *JOAC* conclusion that cross-domain synergy is the best approach to counter–anti-access operations and its lengthy efforts to provide official or doctrinal definitions for the various anti-access–related terms (like most joint documents) do not necessarily provide a basis for programmatic decisions. Presumably that is what the AirSea Battle operational concept is intended to do—explain the detailed logic behind such proposed decisions. A statement by Admiral Greenert that the AirSea Battle Office has come up with "more than 200 initiatives" certainly implies that force structure recommendations are being made, albeit not with the public transparency that an unclassified document could provide.

One source has suggested that some of the "initiatives" have been "operational suggestions" to the combatant commanders on which platforms can be combined in previously unused fashions to create synergy between the two services, although not necessarily in the cross-domain manner prescribed by the *JOAC*. Reportedly Navy P-3 Orion aircraft and U.S. Air Force tank-killing A-10 Warthogs were used in tandem to destroy maritime targets in operations in support of the Libyan revolution, certainly an unaccustomed use of the A-10. In this, the source suggests that—although having no role in the Libya operation—the AirSea Battle Office would function in the future as "a help desk for 21st Century warfare." In other words, the office is not intended to develop a comprehensive approach to countering anti-access warfare, but to develop proposals on how to fill the gaps between current capabilities, procedures, and counter–anti-access requirements.[52]

CSBA AirSea Battle Initiatives

Meanwhile, starting in 2010 CSBA developed several studies intended to inspire and, in fact, propose comprehensive AirSea Battle approaches to countering anti-access warfare, with a willingness to discuss potential opponents and their capabilities. First was Andrew Krepinevich's *Why AirSea Battle?*, which provides a post facto justification for AirSea Battle concept development and, more important, a concise summary of PRC and Iranian anti-access capabilities. Krepinevich urges that the DOD AirSea Battle initiative take a comprehensive and transparent approach on how U.S. forces can deal with the real-world anti-access threats and whether changes in doctrine and force structure are imperative. In his view, "while the effort has to date only been described in the most general of terms, to be effective any new operational concept must be designed to address a particular type of challenge, which forms the object of its design. In this case, the effort should (and appears to) focus on the rising challenge to the US military's power-projection capabilities, which take full expression in China's rapidly developing anti-access/area-denial (A2/AD) capabilities and Iran's similar (albeit far more modest) capabilities, which are focused primarily on the Persian Gulf."[53]

Similar to Krepinevich's earlier work, *Why AirSea Battle?* expresses alarm at what he views as a rapid loss of relative power-projection capabilities of U.S. forces. In this case he includes concern that the loss of access to power-projection forces could have an economic effect, portending loss of access to resources and markets—almost a Mahanian view: "The US military's role as the steward of the global commons has enabled the free movement of goods around the world, facilitating both general peace and prosperity."[54] The need for effective power projection "became even more acute as economic globalization accelerated toward the end of the twentieth century, and, along with it, increased US reliance on global supply chains for a wide variety of goods and services."[55]

Power projection is also the key to being able to assist the defense of allies and partner nations, most of which are located across oceans and closer to potentially hostile states than to the continental United States (CONUS). Krepinevich leans toward the view that the entire U.S. alliance system is dependent on the U.S. power-projection capabilities that reassure allies that the United States is committed and actually can honor its treaty commitments. His view is that this reassurance is in great danger if it is perceived that the United States is unable to penetrate anti-access defenses that extend into the sea areas off Asia or in the Persian Gulf region. Two long-term treaty allies, Japan and South Korea, are located within these sea areas, as are a host of U.S. partners. Although he does not express it, countering anti-access strategies becomes a part of peacetime foreign and economic policies, as well as military posture. As this book argues, peacetime foreign and economic policy and military posture are primary components of grand strategy. Krepinevich describes a future era in which—unable or unwilling to overcome the anti-access challenge—the United States could have to accept disastrous changes in global security. To avoid this, he advises that U.S. decision-makers must not avoid serious assessments and public debate:

> This era of US military dominance is waning at an increasing and alarming rate. Nowhere is this more pronounced than in the United States' ability to project power. Specifically, several states, notably China and Iran, are strenuously working to raise precipitously over time—and perhaps prohibitively—the cost to the United States of projecting power into two areas of vital interest: the Western Pacific and the Persian Gulf. Their efforts present US leaders with a strategic choice of the first magnitude: either acquiesce in the advent of a new world order in which the United States can no longer freely access areas crucial to its economic well-being, or effectively assist key allies and partners in those areas in defending themselves from aggression or coercion. In order to make an informed strategic choice, US leaders will need, at a minimum, to understand the character of the challenges being posed to the country's power-projection forces, and the options available to offset these challenges.[56]

This requires—in Krepinevich's view—that the AirSea Battle be more about doctrinal change than about providing "a help desk." This follows his previous emphasis on establishing joint approaches to what he views as joint problems not just at the level of a joint strategic vision, but down to integrated tactical procedures that might require changes to Service cultures:

> The AirSea Battle concept requires both Services to identify key mission areas and tasks in which their units would operate in an integrated fashion. Then dual-Service tactics and procedures—the operational "nuts and bolts"—would need to be established to enable

> the effective execution of these missions and tasks, and incorporated into dual-Service doctrine. Once established as doctrine, recurrent dual-Service training and routine exercising of integrated missions and operations must become the norm. Much of the above would go against the institutional grain of each Service, since the impact on Service culture and ethos could be considerable. Thus change of this kind could well take a long time to implement.[57]

Whether or not one accepts the need for significant changes to service cultures, it is quite apparent that Krepinevich is advocating a deep, overt, and all-encompassing look across program requirements, acquisition choices, and doctrinal methodology, specifically for the preservation of robust power-projection capabilities in the face of evolving anti-access capabilities. Thus far, this is not something that the AirSea Battle effort has publicly provided.

It is, however, something CSBA *has* endeavored to provide. Later in 2010 CSBA also released a report principally authored by retired Navy captain Jan Van Tol, another Net Assessment alumnus, that—as indicated in its subtitle—details a "point of departure operational concept" for AirSea Battle. This it does by providing the document that many expected an AirSea Battle white paper to be: an extensive evaluation of PRC anti-access capabilities and a recommended list of changes and improvement to current U.S. military capabilities that could help to deter or defeat PRC anti-access activities. Although devoted to the western Pacific theater of operations, Van Tol's report provides a detailed framework for the military components needed to conduct any modern counter–anti-access campaign, a point that the principal author acknowledges. In this way it provides additional illustrative detail to the overall anti-access concept. If Krepinevich's 2010 report is intended to identify the problem, Van Tol's is intended as an outline of the solution. Many of his observations will be discussed later in this book, but it is possible to illustrate the scope of the work by citing its summary recommendations for transforming the unclear image of AirSea Battle into determined defense planning with an effect on future programmatic choices:

> Neither the Defense Department's Program of Record forces and modernization profile, nor current Air Force and Navy concepts of operations accord sufficient weight to the capabilities needed to execute an AirSea Battle campaign successfully along the lines of the one

described in this report. This report recommends multiple initiatives the Air Force and the Navy should undertake, mostly on a dual-Service basis, to field the necessary forces and capabilities for AirSea Battle. These include initiatives on:

- Mitigating the missile threat to Guam and other selected bases, and to maritime forces;
- Correcting the PLA-US imbalance in long-range strike for high-value and/or time-sensitive targets, to include developing and fielding greater penetrating and stand-off long-range ISR and precision strike capabilities and capacities;
- Enhancing capabilities for undersea operations, to include submarines, submersible robotic systems, and mines;
- Offsetting the vulnerabilities of space-based C^2, communications, and ISR capabilities and capacities, to include fielding high-capacity airborne C^3 relay networks to back up space-based systems;
- Emphasizing future standardization and interoperability of data links, data structures, and C^2 and ISR infrastructures;
- Increasing emphasis on and investment in cross-Service electronic warfare capabilities and capacities;
- Enhancing cyberwarfare offensive and defensive capabilities; and
- Developing and fielding directed-energy weapons.[58]

The appropriateness, effectiveness, and affordability of each individual recommendation for U.S. defense policy can be debated. But for the purpose of understanding the requirements for a modern counter–anti-access campaign, the study is a valid guide. In 2012 CSBA released a similarly detailed report by retired Air Force colonel Mark Gunzinger (with Chris Dougherty) titled *Outside-In: Operating from Range to Defeat Iran's Anti-Access and Area Denial Threat*, outlining a similar counter–anti-access campaign in Southwest Asia.

Relationship to Asymmetrical Warfare

Another concept that developed in the past two decades and gained considerable traction among defense analysts and in the DOD is asymmetrical warfare (also "asymmetric" in the literature). This too was largely a result of Desert Storm but also of the disproportionate size of the U.S. defense investment in comparison to the other countries of the world. With the fall of the Soviet Union, the United States became the sole superpower—which in military terms could be defined as the only nation that could project military power on a global basis in a comprehensive and sustained fashion. Other nations, such as the United Kingdom and France, could project limited military power on a global basis. They, however, are U.S. allies with similar interests. NATO could do this as a collective but largely based on U.S. logistical assets. Others, such as Australia, could project power within their immediate regions but not globally. Even the much-reduced Russian military could somewhat haltingly project power within its "near abroad."

Since potential opponents could not afford the same quality or quantity of military forces as the United States, it was perceived that they would most likely attempt to fight using asymmetrical techniques. At its simplest, asymmetrical warfare means the utilization of weapons or tactics that are different than those used by well-equipped conventional forces with a wide portfolio of assets. Some of these techniques might not be in accordance with the generally accepted international laws of war that were codified to reduce the suffering of noncombatants, civilian populations, and prisoners of war, and that prohibit the use of weapons considered particularly heinous, such as chemical or biological weapons. As Roger Barnett notes, simple or "true" asymmetry constitutes "those actions that an adversary can exercise that you either cannot or will not."[59]

Asymmetrical warfare may be the logical approach of a weaker opponent confronting a stronger, but it is not necessarily confined to use by the weaker. It is a relative term based on perspective. Capabilities of U.S. forces could certainly be perceived as asymmetrical by the more modest and less capable forces of other states that may not be able to operate in certain domains or mediums. Asymmetrical techniques could be used by forces that are relatively equal. Rod Thornton makes this point that asymmetrical "does not mean unequal. 'Symmetrical' implies a mirror image; sometimes that image can be smaller, but nonetheless a likeness exists. 'Asymmetrical' implies a relationship that cannot be considered to be alike. If one side in a conflict, for example, has lots of tanks and its opponent has far fewer, then the battle would still be symmetrical."[60]

Asymmetrical warfare can involve weapon systems developed through sidewise technologies or the use of "legacy" techniques such as cavalry. But it can also utilize types of warfare—or perhaps we should call it violence—that Western states (at least overtly) do not use, such as terrorist attacks deliberately targeted at unarmed civilians by nonstate actors or proxies.

Asymmetrical techniques can be used in conflicts that are not inherently anti-access. Insurgent guerrilla warfare is usually targeted against one's own government. Insurgent actions could be seen to consist of area denial in that the insurgent force attempts to deny the government freedom of action—even in regions ostensibly under government control. But unless an outside power becomes involved, this does not necessarily involve prevention of the entry of forces from outside the region.

From this perspective, asymmetrical warfare cannot be seen as a subset of anti-access warfare. Nor do anti-access operations necessarily involve asymmetrical warfare. As an example similar to Thornton's, a military can possess fewer satellite-based sensors than its opponent and use them as part of an anti-access strategy, but this would be a symmetrical technique as regards an opponent with many more satellites that utilizes its information to conduct counter–anti-access operations. However, asymmetrical techniques *can* be used as part of an overall anti-access effort, either replacing or supplementing conventional operations. Harking back to classical Chinese writings on strategy, PLA commentators have stressed asymmetrical approaches in which "the inferior can defeat the superior."[61]

If envisioned as a Venn diagram, anti-access warfare and asymmetrical warfare can be seen as two circles that intersect and share a variety of techniques as well as a set of logical assumptions. But neither is dependent on the other. In assessing anti-access strategies, one must assume that the anti-access force will use both asymmetrical and symmetrical techniques. The counter–anti-access force can do so as well.

In addition to asymmetries of techniques, forces, and weapon systems, there is also a possible asymmetry in objectives that is important to an understanding of the anti-access concept. This particularly comes into focus through the lens of the fifth fundamental element, the impact of extrinsic events. Even when defending conquered territory, the anti-access state is likely to perceive that it is fighting for survival of its current regime. The out-of-area counter–anti-access power may be taking action to defend an ally or preserve global stability, but it generally does not perceive itself as pursuing a national survival-level objective. In discussing the RMA, Krepinevich notes that "asymmetry . . . will likely exist in terms of competitor goals and objectives. If we retain our Cold War era objective of being a global power, we will find that the military-technical needs of many competitors, whose ambitions are regional, are likely to be [different]. . . . National objectives will help define a competitor's requirements, and its level of success as well."[62] This is true of the determination of the competitor as well, which is why some anti-access states have clung to the idea that causing significant casualties to U.S. forces could persuade U.S. decision-makers to cease counter–anti-access operations. After all, runs the logic, the outcome is not as important to the United States, an extraregional power, as it is to the anti-access state, which is willing to absorb greater casualties. American "casualty aversion" is thus seen as an asymmetrical vulnerability to be exploited by anti-access forces, and the maximization of enemy casualties may become an objective of the anti-access efforts.[63]

Whatever the truth of American casualty aversion, we have already observed a historical basis for the asymmetry of objectives. The Greeks were fighting for their survival; Xerxes was fighting for revenge and to add yet another piece of territory to his already large empire. Giving up his efforts did not have the same direct costs as giving up would have for the Greeks, and potential extrinsic events convinced him that the cost to continue was too great and the cost to quitting relatively low.

Anti-Access Networks

Robert Work's emphasis on describing anti-access efforts as a network has resonance. To be effective, an anti-access strategy must coordinate its elements so that they work together in order to prevent any obvious gap or seam that the enemy can exploit. If the wall has gaps, the enemy can exploit them. Modern weapon systems must operate as networks. In order to target an enemy over the horizon, spaceborne or airborne sensors must be used for detection and targeting information that are sent to offensive or defensive weapon systems so they can be successfully employed. Over-the-horizon backscatter (OTH-B) radars and other techniques utilizing sky waves can also provide long-range information, but most earthbound radars are limited in surface range by the curvature of the earth. Whatever the source of information, it must be processed and analyzed so that the right targets can be identified and tracked, and their future locations predicted. All this requires a network along which information flows from sensor to shooter. Postattack information—known as battle damage assessment (BDA)—also needs to flow in order to determine whether the attack has been successful or the enemy needs to be reengaged. The information is also vital in the choice of which weapon systems or types of ordnance are to be used, exchanges or "hand-offs" between firing units, and coordination of dispersed platforms. All these operations are complex; sometimes complexity can be the enemy of effectiveness.

Targeting particular aspects or nodes of the network might be able to bring the whole complex process to a halt. Without sensor information, weapons can only be shot blindly. In a saturation attack, in which the shooter fires a considerable number of weapons at the probable location of the enemy, there is always the chance of lucky hits. But if the enemy has been using deceptive techniques, it is possible that all the weapons aim at the wrong location. Unless the shooter's supply of ammunition is unlimited—and it never is—it is possible to run out without doing much damage to the enemy even if the enemy cannot effectively attack the firing units. Without sensor information and with only preset coordinates, Saddam Hussein's forces fired off scores of Scud missiles (admittedly never very accurate to begin with) during Operation Desert Storm but only a couple had any military effect, and the politico-psychological effect (an attempt to provoke the involvement of Israel) was successfully mitigated by the coalition. The point is that viewing the anti-access efforts as a network of strategies, techniques, and systems prompts the understanding that defeat of a particular portion of the network could seriously degrade it. Likewise, it points to the need for redundancy in systems and fallback strategies in order to deal with critical damage.

Access and Forward Presence

A concept that needs to be discussed in terms of access, anti-access, operational access, and assured access is that of forward presence. This is a naval term that seeks to describe the ability—and regular practice—of combat-ready naval forces to be routinely positioned in potential regions of crises during times of peace or nonhostility. Forward presence implies the continual presence of combat-ready forces to areas where the United States does not have fixed land bases, often just offshore inherently hostile states that might not otherwise be deterred from precipitous actions against neighbors. During the 1990s, and even before, forward presence was considered one of four major strategic concepts or missions of the Navy and Marine Corps that indicate their worth to the nation in times of peace.[64] Since the Law of the Sea allows the unfettered access of military ships to the high seas up to twelve nautical miles off the coast of any sovereign state—a right the PRC frequently contests—it is possible for naval forces to provide effective forward presence without the necessity of land bases or the permission of an ally or partner. An aircraft carrier task group parked just offshore can constitute quite a presence and is presumably an effective deterrent to the outbreak of local hostilities.

For most of its existence, Army and Air Force planners have hated the concept of forward presence, since they saw it as a largely effective way for the Department of the Navy to justify its peacetime force structure.[65] This changed somewhat in the late 1990s when both of the others services attempted to redefine themselves as expeditionary forces not tied to extensive basing requirements. In the competition for resource justifications, the Air Force went so far as to count every landing of any aircraft in a foreign airfield as a forward-presence mission, and the Army argued that its overseas liaison officers and attachés constituted forward presence.

With the advent of the anti-access concept, and particularly with the development of more sophisticated weaponry by the PRC—ASBMs being the poster child—many analysts have suggested that the naval forward-presence mission and indeed the joint forward-presence mission, if such exists, are no longer viable. CSBA studies have frequently cited relatively unsophisticated (and sophisticated) guided rockets, artillery, mortars, and missiles (under the acronym G-RAMMs) with increasing ranges as constituting a severe threat to naval platforms, particularly to amphibious forces. Naval-favoring analysts have argued that fewer nations than expected are developing such anti-access capabilities, but with a defense community focus on Iran and the PRC, this has not had much of an effect on critics of naval forward presence.[66]

In any event, the ability of naval forward presence to provide deterrence against states with extensive anti-access networks and capabilities *is* a meaningful question. During their tenures as CNO in the 1990s, Adm. Jeremy M. Boorda and Adm. Jay Johnson argued that the Navy could provide assured access to regions of potential conflict for the joint force, given modest advancements in counter–anti-access capabilities. Many naval platforms, such as *Arleigh Burke*–class guided-missile destroyers, have what are considered quite good defenses against cruise missiles and, in certain ships, theater ballistic missiles, and according to unclassified information U.S. nuclear-powered submarines are greatly immune to all but the most sophisticated antisubmarines efforts. But with the decision to procure the much less survivable LCSs and base them overseas, the trend appears to be away from robust presence as deterrence in anti-access environments—a trend that Robert Work noted as scholar, if not as undersecretary.[67]

Post–Cold War Assumption of Access

A major reason anti-access warfare is considered such a significant challenge to the joint force and countering it a mission of the overall Department of Defense was the "assumption of access" that settled into U.S. military planning in the immediate post–Cold War world. Many of the sources previously discussed have identified this perhaps unwarranted assumption of access as leading to the development of service acquisition programs and joint force structure that could be put at risk by opponents' increasing anti-access capabilities. These observations have particular weight in an acquisition environment in which it takes the DOD at least a decade to field a new major combat system.

Partly because of assumption of oceanic sanctuary and the security of overseas land bases, but even more because of a desire for efficiency and cost savings, the DOD has built significant vulnerabilities into its systems and operational procedures. These are not just physical security vulnerabilities, but involve the viability of the high-technology systems and networks that are defining the modern American way of war. Whether or not the risks were initially recognized, these vulnerabilities can be particularly exploited by a well-planned anti-access strategy. U.S. forces have become greatly dependent on satellite-based communications and sensors, which—with the collapse of the Soviet Union's military space programs—appeared immune to a general attack. Yet space systems are logically the initial target of an effective anti-access network of weapon systems. Internet connectivity is perhaps an even greater vulnerability. PRC commentators have repeatedly identified U.S. reliance on space systems and computer networks as the most significant vulnerability.[68] Van Tol provides an excellent discussion of the effects of the assumption of access that was prevalent in the previous two decades and—quite frankly—resulted from shortsighted or expedient decisions by the DOD:

> US ground, air and naval forces have long been accustomed to operating from sanctuary. . . . US communications, ISR, and precision-guided munitions (PGM) are heavily dependent on high-bandwidth connectivity for command and control, target detection, precision strike, and post-strike battle damage assessment operations. This connectivity is highly reliant on long-haul space-based assets that have hitherto also been accorded sanctuary status, save for the occasional modest localized jamming. The same can be said with respect to cyberspace which, despite numerous and consistent probes by China and other states, and by nonstate entities and individuals, has never been seriously compromised. The growing Chinese A2/AD capabilities, to include its cyberweapons, threaten to violate these long-standing sanctuaries. As this occurs, the consequences for US forces would include:
>
> - Loss of forward sanctuaries in physical domains and virtual domains (including space, cyberspace, and the electromagnetic spectrum);
> - Denial of access to areas of operations; and consequently
> - Loss of strategic and operational initiative.[69]

The development of weapon systems that can support viable high-technology anti-access strategies turns the assumption of access on its head. An assumption that access will *not* be bitterly contested by potential opponents becomes a weapon in the hands of a perceptions-manipulating opponent. The ultimate anti-access strategy—idealized and nearly impossible to implement in the information age—would be for the anti-access state to allow the strategically superior power to believe that it can still achieve access at minimal cost to its forces. If hostilities were to occur, the reverse "shock and awe" (to use that largely discredited term) might cause the decision-makers of the strategically superior state to conclude the intervention is just not worth it. This stratagem is in accordance with the fourth fundamental element, the criticality of information and intelligence and the effects of deception. One of the purposes of this book, along with the other published studies on anti-access, is to ensure that such does not occur.

Basing and Access

Overseas bases are certainly assets as well as vulnerabilities. A base utilized by U.S. forces on the soil of another sovereign nation is a strong symbol of commitment by the United States to the defense of the host nation and a commitment by the host nation to its alliance with the United States. There are always a variety of local issues involving base use and the impact on communities (just as there are such issues concerning bases in the United States), but such bases would not exist without the voluntary cooperation of the host nation. There are a number of cases where the host nation has requested the removal of U.S. forces from a particular base or entire territory, and the United States has always complied. In any event, the stationing of U.S. forces within another nation builds, and is indicative of, a strong relationship, one that is designed to have a deterrent effect on hostile neighbors.

Overseas bases are also closer to the likely crisis areas of the world, as well as potential opponents of the United States. Like the U.S. need for power-projection forces and the attractiveness of utilizing anti-access strategies against the entry of U.S. forces, their positioning resulted from the facts of geography and history that have provided the United States with peaceful neighbors and two oceanic moats. It has been logical for the United States to rely on overseas bases for its defense posture in the past, either because they were not perceived to be vulnerable to forces in the immediate region or because they functioned as Cold War trip wires by which an attack against them would unleash a potential global conflict.

The sources we have examined in defining the outlines of the modern concept of anti-access have all pointed to fixed overseas bases as greatly vulnerable. There is *no* contemporary source—even among those advocating that the DOD fix its primary attention on issues other than anti-access—that does not acknowledge the growing physical vulnerability of fixed land bases. Many of the existing anti-access studies recommend specific measures to reduce this vulnerability, although even the most optimistic admit that it cannot be completely eliminated against an opponent with the capability to conduct multiple strikes against fixed coordinates. This issue is plain.

But there needs to be acknowledgement that there is also a parochial issue that surrounds the debate concerning the vulnerability of fixed bases, one that often clouds dispassionate examination of all options to mitigate the vulnerability. The U.S. Air Force was created out of the Army Air Corps. Naturally enough, its tactical aviation flies from land bases, with those land bases needing to be within operational range of any opponent. Midair tanking can extend the range of combat aircraft, but it is logistically difficult and is itself a vulnerability in an anti-access environment. As Bowie's RAND study points out, recent programmatic decisions have shifted the ratio between fighter/attack aircraft, with their inherently short ranges, and long-range bombers. Meanwhile, sea-based aviation remains part of the U.S. Navy and is treated as an organic, integrated combat capability of the maritime domain. In fact, more sailors are assigned to naval aviation commands (including aircraft carriers) than to the other surface ships and submarines. A shift in emphasis—or more important, resources—in attack aviation from fixed land bases to mobile sea bases (i.e., aircraft carriers) has been traditionally viewed by the Air Force as a threat to its core mission and to its self-perception as the service that should control the air domain of the joint forces environment. As previously discussed, the current concept of jointness imposes an orthodoxy of nearly equal participation in all missions. Within this culture, there is little likelihood of blank-sheet analysis of the costs—in strategy as well as matériel—of mitigating the vulnerability of fixed land bases and mobile sea bases in an anti-access environment. It is unlikely that such can be facilitated under the AirSea Battle concept; it certainly would not be seen as a cooperative effort. At the same time, the range of carrier-based attack aircraft has actually shrunk since the Cold War—through the Navy's deliberate choices—and the potential vulnerability of aircraft carriers has appeared to increase.

Adding another dilemma to the vulnerability of fixed land bases is the option of shifting to operations primarily from air bases located in

U.S. territory—the premise of a concept called Prompt Global Strike (PGS). However, this could potentially shift the focus of anti-access forces away from regional conflict and toward preemptive conventional strikes on CONUS itself.

Limitations to the Concept of Area Denial

Earlier this book states that the linkage between anti-access and area denial is an artificial construct. It is appropriate to discuss this in terms of the development of anti-access as a strategy.

An anti-access strategy is designed to prevent the entry of power-projection forces into a contested region and to eliminate any such forces that might already be there. It is the antithesis of engaging forces that might be overwhelming while operating on land within the heart of the region. Area denial as defined by the *JOAC* and other sources consists of tactics and systems to be utilized in denying the freedom of action of forces operating within the region. In other words, the tactics of area denial avoid force-on-force combat—but against an enemy that is already there. Anti-access and area denial appear to be very different situations.

Moreover, the operations involved in area denial—indirect and harassing fires, operational maneuver, deception, strike-move-hide—appear to be techniques that are historically common to all ground combat, particularly by a weaker opponent. It could be assumed that guerrilla warfare is primarily area denial. Therefore, it is difficult to determine a difference between area denial and ground combat as it is traditionally fought in unequal contests.

It is possible to argue that anti-access and area denial should be linked because the weapon systems utilized are the same or similar but shorter in range. In fact, the *JOAC*, looking at anti-access as simply a military problem to be solved—an opponent's campaign to be defeated—certainly implies that, positing in the very definition of anti-access that it is usually conducted at long range. But this is not a method capable of assessing anti-access as a war-fighting strategy (as opposed to operational technique) and element of grand strategy.

It is also possible to argue that anti-access and area denial must be linked in order to preserve equanimity among all services and support the current conception of jointness.

Integrating the Other Elements of Power

Understanding the strategy of anti-access, as opposed to simply assessing anti-access weaponry and tactics, requires one to integrate other elements of international power into the concept. As has been seen, diplomacy or economic relations can pose a threat to the existence of oversea bases if the anti-access state can persuade the host nation not to allow their use by a counter–anti-access power. The ability to project power is thereby inhibited without any clash of forces. Diplomatic pressure toward changing the Law of the Sea to prohibit military vessels from a nation's claimed two-hundred-nautical-mile economic exclusion zone—an obvious objective of the PRC—would also inhibit power projection, particularly in a relatively narrow sea area such as in East Asia. Such efforts not only support an anti-access strategy, but are integral to it, which is why PRC sources often discuss the "three warfares" of psychological, media, and legal warfare.[70] At the same time, the power projector can use diplomatic and economic relations to ensure that host nations maintain their support, as well as mutual defense efforts to reduce pressure of coercion by the anti-access state.

Diplomatic and economic efforts can be used to encourage the extrinsic events that influence the counter–anti-access power to shift its focus onto other, perhaps even more pressing problems. This has been a routine effort of historical anti-access campaigns. Commentators have argued that in recent years, America's fixation on events in the Middle East has blinded it to even more dangerous threats elsewhere. That is not to imply that Middle Eastern events have been shaped elsewhere—only that the attention of even powerful states can be limited in scope. But again, diplomacy might be used to stifle extrinsic events, quieting the international environment and focusing world support for the counter–anti-access effort. The United States was able to do that in World War II and in fashioning a coalition for Desert Storm.

The United States, the nation with the most capable power-projection forces and potentially robust counter–anti-access capabilities, is said to also possess the greatest amount of soft power through the attractiveness of its form of government, its culture, and its political and economic freedom. One might ask if soft power has any role to play in the concept of anti-access or as part of a counter–anti-access campaign. The answer lies in the time frame one uses in assessing counter–anti-access. In a short time frame, and thinking strictly in terms of military operations, it is hard to see how soft power makes any impact. It is difficult, if not impossible, to operationalize soft power—particularly by a democracy that allows for freedom of belief and expression. But if one examines long-term anti-access strategies in terms of the construction of "great walls" surrounding societies and preventing their interactions with outside influences, soft power could play a role, even without any action by a state, in chipping away at the battlements. It might be even possible to enhance this role through the judicious use of media and psychological operations.

In determining how other elements of international power could be utilized effectively, the objectives of the states adopting the anti-access and counter–anti-access postures must also be assessed. In the case of Imperial Japan, an anti-access posture was adopted in order to protect its gains and prospective additional gains of conquest. The target of the anti-access posture, the United States and its allies, could either choose to conduct a counter–anti-access campaign or acquiesce to Imperial Japan's objectives. America's counter–anti-access potential had failed to deter Imperial Japan's actions. In the current situation in which the flash point in any conflict between the PRC and the United States would be forcible conquest of Taiwan, it is possible that U.S. counter–anti-access capabilities could have a deterrent effect on PRC actions, since PRC anti-access efforts seem to be initially aimed at preventing the United States from intervening in the Taiwan scenario. Adding diplomatic efforts and managing economic relations could enhance the deterrent effect of the counter–anti-access capabilities.

If diplomatic and economic relations are to be included in a counter–anti-access strategy, particularly in a government in which the military is subordinate to civilian control and acts as but a part of the government, deterrence or defeat of a potential opponent's anti-access strategy must be more than a military problem. Using current buzzwords, it needs a "whole of government" approach to its solution. But whether that is possible in a society that has conflicting opinions and interests is difficult to say. Perhaps it depends on how clear and present the danger from the anti-access state appears to the majority.

Anti-Access, Counter–Anti-Access, and Grand Strategy

Grand strategy has always been difficult to define. Many studies of historical grand strategies have not even attempted to give it a specific definition. As in the proverbial Supreme Court decision, you know it when you see it. During the Cold War, the grand strategy of the United States was to preserve its own independence, freedom of action, and economic growth while containing the Soviet Union until the internal contradictions in communism brought about its own demise.

For the purpose of this book, we will assume that grand strategy entails all actions intended to preserve a state's (or a regime's) freedom of action in order to achieve the objectives it desires. If a state or a regime perceives that its objectives are in conflict with a strategically superior power and this conflict might lead to armed hostilities, a military anti-access posture would be a useful part of its grand strategy. One of the reasons that democratic states do not adopt an anti-access posture toward other democratic states, even if the international environment is theoretically always in a state of anarchy, is that they do not perceive themselves to be in a position of hostility. The European Union may have a host of political or economic issues with the United States, but it expects negotiations and compromise, not hostilities, to be the result. That is why anti-access postures are generally the province of authoritarian states. For a state or regime whose grand strategy revolves around the consolidation or retention of internal power in a world in which external powers might be hostile to it, building a great wall consisting of an anti-access strategy would also be desirable.

The emphasis on grand strategy generates an emphasis on deterrence in turn. John M. Collins, author of the most comprehensive book on military strategy as an element of grand strategy, explains this emphasis as a defining characteristic of grand strategy.[71]

If an anti-access strategy is adopted by a state, it should be seen as an element intended to support its grand strategy. Perhaps the best way to defeat an anti-access strategy is to be able to hold the objectives of the grand strategy at credible risk.

The Breaking of Great Walls

Returning to our metaphor, when a wall is broken into rubble, the component pieces—brick, stone, whatever—can no longer perform their designed function. They may still have the ability to harm; one can stub or even break one's toe on an isolated stone. But lying apart, they can no longer stem the flow of wind, water, or force rushing through the gap and expanding into the interior.

Building an anti-access strategy is very much like building a wall. One places political, diplomatic, and economic initiatives alongside military systems, doctrine, and tactics. In the operational military effort, one tries to develop and emplace defensive and offensive weaponry in all mediums or domains—maritime, air, land, space, cyberspace—and ensure that they can work together, so that, unlike Xerxes at Thermopylae, the counter–anti-access forces cannot find a path by which the determined defense can be flanked.

In the opposite direction, degrading or destroying anti-access strategies are like breaking down a wall. One can do it methodically, piece by piece. But one can also strike at the base or most vulnerable point in order to make the bricks above crash down.

Speaking of striking the most vulnerable points evokes the idea of a center of gravity, or even the image of concentric rings from the writings of retired Air Force colonel John Warden.[72] But the anti-access network itself is not the center of gravity for an opponent. The anti-access network is simply the system by which the center of gravity is protected. In the constructs of Warden and others, the enemy's defenses should be bypassed in order to strike at the center; resources should be focused on the center, not on taking apart the defenses. But an effective anti-access network is not one that can be bypassed. It must be fought until it is ineffective. If it could be bypassed from the outset, it was simply never effective.

The purpose of the *JOAC* and the many unofficial studies of anti-access is to assess and prepare to counter anti-access networks that cannot be bypassed from the outset. Being able to bypass defenses from the outset means that one *does* have a degree of assured access. This does not mean that countering anti-access will always be completely an attrition-based campaign. It will require maneuver and a variety of other operational techniques. But it is likely that a lot of attrition must occur until the foundation is weakened so much that a breakthrough can occur.

An important point is that in developing a counter–anti-access campaign, one might be able to strike the key points to break the wall, but that does not necessarily bring any end to the conflict. The power-projection phase must occur, and the opponent must be defeated or his motivating objective denied.

The counter–anti-access campaign may therefore be protracted. It may not have to occur at all if it deters opponents from their objectives. It may even involve the penetration of ideas and encouragement rather than weapons, even while weapons are necessary to establish a deterrent effect.

The remainder of this book will not discuss the means, legality, or morality of encouraging, assisting, or fomenting revolutions in authoritarian states as a technique of countering anti-access defenses. These details will deliberately remain outside the scope of the present book. Yet the use of internal—perhaps we can call them intrinsic events—as a means of countering anti-access strategies is as logical as the encouragement of extrinsic events by the anti-access state in order to interfere with a strategically superior power's attention, decision-making, and/or commitment.

Such internal events pose a primary threat to the anti-access walls of an authoritarian state in the same way that frost heave poses the primary threat to the many stone walls built in New England during the period when most of the land was under cultivation. Frost heave has knocked down more stone walls than later inhabitants have carried away.

However, unlike frost heave, internal events can be encouraged. This must be acknowledged as a possible technique of counter–anti-access operations. This also points to the richness and usefulness of understanding the concept of anti-access as a strategy.

Chapter 3

The Anti-Access Campaign and Its Defeat

A concept does not a strategy make. A concept is merely an analytical sketch, a means of defining the outlines of an idea or phenomenon.

Definitions for the word *strategy* are many, but perhaps the simplest comes from another scholar-warrior, late Rear Adm. J. C. Wylie, USN: "Strategy is a plan of action designed in order to achieve some end; a purpose together with a system of measures for its accomplishment."[1] An anti-access strategy is a plan to prevent an opponent from operating military forces near, into, or within a contested region. This plan, if it is to support the objectives of grand strategy, includes links to other plans of action for achieving other, usually related objectives.

An *anti-access campaign* consists of the implementation of such a plan. The ultimate measure of its accomplishment is the continued exclusion of a superior opposing force from the contested region until time, attrition, and/or extrinsic events shake the determination of the superior force. Its objective is *not* to achieve victory in a symmetrical force-on-force battle. Rather, its purpose includes the avoidance of such a direct force-on-force contest within the region. In support of other objectives already achieved, a stalemate can be a victory for the anti-access force.

A *counter–anti-access campaign* attempts to defeat an anti-access strategy and achieve operational access. But for the counter–anti-access force, operational access is neither the final objective nor measure of full accomplishment. The military objective is to defeat the anti-access effort designed to keep one out of the region and then defeat the enemy from within the region. From this perspective, a counter–anti-access campaign is an inherently more difficult effort than the anti-access effort. Countering anti-access is but the first phase in an effort of two (or more) phases. This was evident in the war against Imperial Japan. First the United States and its allies had to be able to enter in force into the Japanese-held island region—in other words, gain access to the region. Once access was assured, they then had to destroy the opposing Japanese force, reverse their gains, eliminate Japanese commerce, and prepare to invade the Japanese homeland itself. The last task was avoided through the use of the atomic bombs.

This effort of two (or more) phases parallels how most naval strategists view the role of navies in war on land and sea. The first phase consists of defeat of the enemy's naval forces—the war at sea. This is the defeat of an anti-access effort within the maritime domain, dimension, or medium. The second phase, once sea control or command of the sea has been achieved, is the projection of power ashore and support of combat on and over the land. Capt. Dudley Knox, USN, a naval strategist of the 1920s and 1930s, expressed this two-phase view with the following admonition: "The supreme test of the naval strategist is the depth of his comprehension of the intimate relationship between sea power and land power, and of the truth that basically all effort afloat should be directed at an effect ashore."[2] Wiley also uses this two-phase explanation of the objectives of naval warfare. This again points to the naval roots of the anti-access concept.

Deterrence: Ultimate Strategic Goal

Before analyzing anti-access and counter–anti-access campaigns, it is useful to discuss the precampaign phase, what recently has been identified in defense discussions as *phase 0*. We will refer to it as the *deterrence phase*. Returning from the level of campaign—expressed in doctrine as the operational level of war—to the level of strategy, we must again note that the ultimate goal of a viable strategy is to avoid having to implement it as a campaign. Viable in this sense means a strategy combined with the actual forces necessary to successfully implement it. Military forces, including anti-access or counter–anti-access forces, are designed to be effective in combat but are most effective by achieving the objective while not actually being used in combat. This avoidance is logical, although it might also seem paradoxical to those who view as inefficient the spending of money on resources for wars that never occur. A great deal of resources must often be spent to ensure credible deterrence, yet the forces themselves may never be used. This is not a standard business model.

Perhaps the greatest example of such a victory was the outcome of the Cold War. The U.S. objective was to contain the Soviet Union, ensure no other nations fell under its influence—at least not in Europe—and deter the outbreak of a global war until the Soviet Union and its empire collapsed from within. Scholars may argue about who was most responsible for the end of the Cold War: whether it was American president Ronald Reagan, with his defense buildup and economic pressure on the Soviet Union, or Soviet president Mikhail Gorbachev, with his introduction of *perestroika* and acquiescence to the collapse without resorting to conflict. But it is undeniable that the U.S. strategy succeeded without a war directly between the superpowers.

In similar fashion, the primary or initial purpose of an anti-access posture is not to engage an enemy, but to deter it. By definition, the enemy to be engaged is a strategically and militarily superior power that is likely to prevail in any direct force-on-force conflict. Therefore, avoiding or deterring such a direct attack is the least costly method of successfully engaging the perceived enemy. This correlates with the primary precept of the ancient strategist held in highest regard by modern military thinkers, the probably legendary Chinese sage Sun Tzu, who advised that "to win one hundred victories in one hundred battles is not the acme of skill. To subdue the enemy without fighting is the acme of skill."[3]

Again, this is not simple deterrence of outside attack, a natural posture of any nation-state. Most states that opt for a distinct anti-access approach view a conflict as inevitable. Many intend or are engaged in aggressive military action, such as Imperial Japan in the 1920s and 1930s, and their choice for an anti-access approach is an effort to dissuade or stalemate an outside power capable of limiting or reversing their military gains. Today the states that appear to be adopting anti-access strategies, which include the Islamic Republic of Iran and the People's Republic of China, are perceived by their neighbors as posing a potential military threat, have demonstrated a degree of political aggressiveness concerning regional issues, and are opposed to any interference by the one power with the greatest potential for interfering with their freedom of action, the United States. Worse yet would be facing a coalition of the United States, its allies, and its partner nations, a combination that needs to be prevented or deterred.

Similar in nature, the potential ability of a strategically superior power to conduct a successful counter–anti-access campaign can prove an effective deterrent against aggression. If the anti-access network of an aggressive state is overcome, not only is it likely that its objective is denied, but the survival of its regime is put in jeopardy. At the very least its position in the international environment will be greatly altered, its pretense to power shattered. It may become an international pariah without allies and constrained in the actions in its internal sovereignty—the position of Saddam Hussein's Iraq following the first Gulf War. The aggressive state must calculate the probability that its anti-access network can withstand such an onslaught. Just as certain types of sensors and weapon systems are necessary in building a modern anti-access network, certain sensors and systems are needed to overcome it. If the strategically superior power possesses these systems, its deterrent potential rises.

In discussing deterrence as an element of anti-access or counter–anti-access strategies, one must remember that deterrence is not a physical property—it is a state within the mind. Deterrence occurs when national decision-makers perceive that a potential opponent has the capability and willingness to deny them an objective. Both capability and willingness must be apparent to these decision-makers if they are to be deterred from taking action. Additionally, culture and political environment have a significant effect on decision-making, and any deterrence effort must take into the consideration the cultural and political environment of the decision-makers. In assessing the probable effectiveness of a deterrence posture, cultural mirror-imaging must be avoided. What deters one particular opponent may have less effect on another. Strategic superiority and overwhelming force usually possess strong deterrent potential, but only if the target of deterrence believes they will be used and that the impact of their use would result in a negative outcome within the context of its culture. Returning to the Gulf War of 1990–1991, the political culture of violence in which Saddam Hussein flourished and his belief that the Western nations would not actually use their overwhelming force made him more difficult to deter than was perceived by U.S. diplomats and decision-makers.

If implementation of an anti-access campaign in support of an aggressive objective—the forcible annexation of Taiwan as example—is to be deterred, the anti-access state must be convinced that the counter–anti-access/power-projection state has the capability of defeating or significantly penetrating the anti-access network, that it has the will to do so in the particular situation, and that the cost of defeat is not worth the risk.

Elements or Steps of Deterrence

Deterrence is thus a two-way street. The anti-access state seeks to deter any interference with its objectives. The counter–anti-access power seeks to deter the anti-access state from its objective by making clear that its anti-access effort will fail. There are identifiable elements or steps involved in these efforts.

For the anti-access state, a first step in deterring the involvement of an outside power, or preparing the battlefield if conflict is inevitable, is to conduct political, diplomatic, and economic efforts to prevent other states within the region from achieving or maintaining an effective politico-military relationship with an outside power. During peacetime or, perhaps more properly, a time period without active hostilities, diplomacy and economic incentives are primary tools to foster cooperation, or at least toleration or acquiescence, by other states within the region or beyond. Threats of coercion can also be used, but these can be risky because they can have the effect of driving other states into a firmer alliance with the outside power. The objective in this stage of the anti-access effort is to deter or limit any support of the outside power by other regional states. Most important is to prevent the outside power from being able to station or position military forces within the region. However, any degree of support that can be eliminated—for example, hostile international public opinion—is a useful target for this effort.

This is not easy to achieve for at least two distinct reasons. First of all, the strategically superior power is likely to have achieved its status through its own skillful use of diplomatic and economic relationships, as well as developing its strong military potential. The anti-access state—even if skilled in diplomacy and economic co-optation—finds itself competing with another master of the game. Secondly, the anti-access state must appear to be a lesser security threat to others in the region than the interference of an extraregional power—an appearance that is particularly difficult for an authoritarian state. There is also a natural tendency for smaller or less powerful states to view a regional power with greater suspicion than an extraregional power because it is closer and therefore presents a potentially more immediate threat. In explaining his previously inconceivable support for a tilt toward the United States and away from the Soviet Union, Mao Zedong stated, "Didn't our ancestors counsel negotiating with faraway countries while fighting with those that are near?"[4] Admittedly the PRC could hardly be considered a small state, but as unlikely a source as Mao might be, this is traditional wisdom in many places other than China.

A second step or element is to utilize diplomatic or economic incentives or disincentives to influence the power-projection/counter–anti-access power and convince it that the objectives of the anti-access state are legitimate and strategically insignificant. If an economic dependency can be created in which the counter–anti-access power is beholden to the anti-access state, its involvement might be deterred. If dealing with a democratic state, interest groups could be encouraged to influence governmental action into allowing the anti-access state its objectives. This step can involve a carrot-and-stick approach: promises of improved political or economic relations mixed with threats of a loss of political or economic relations. Threats of cutting off the source of a valuable raw material or access to domestic markets might have some influence. Prolonged negations might delay the counter–anti-access power from reacting quickly to any fait accompli in the region.

A third step or element is to create an environment of international support, or at least acquiescence to the objectives of the anti-access state. Again, the effort might be directed at convincing other nations of the legitimacy of the anti-access state's objectives or providing economic incentives to support its positions. Appeals to international public opinion would certainly be an adjunct to any direct diplomatic efforts. In an anti-access vs. counter–anti-access situation, the potential counter–anti-access state could be portrayed as the aggressor or one that is willing to turn a purely regional dispute into a global war.

TABLE 3-1 Diplomatic and Economic Elements

Diplomatic/Economic Element (Objective)	Techniques	Remarks
Detach regional states from counter–anti-access power	• Reduce appearance of threat • Create economic incentives or disincentives • Conduct media campaign	Coercion might be used but could prompt closer relationship to counter–anti-access power.
Influence counter–anti-access power	• Reduce appearance of threat • Conduct media campaign • Foster demonstrations, "peace movements," scholarly opposition • Lobby business community	Care needs to be used to avoid appearance of coercion or obvious manipulation of possible opponents. Media activities require an understanding of target culture and skill.
Create environment of international support	• Reduce appearance of threat • Create economic incentives or disincentives • Conduct media campaign • Use international forums • Use diasporas	Will be competing with diplomacy and economic relations of counter–anti-access power.

All of these activities could be supported by propaganda, adroit use of media, fostering demonstrations, support for "peace movements," lobbying of the business community associated with its exports, use of international forums, and sponsorship of scholarly opposition. Table 3-1 lists diplomatic and economic elements that can support an anti-access deterrence posture along with supporting activities.

Diplomacy and Economic Relationships with Authoritarian States

Throughout this book we have put emphasis on the attractiveness to authoritarian powers of an anti-access approach. A great part of this attractiveness results from their perception of a hostile environment and fear that other powers naturally seek to thwart their objectives. Such states inevitably carry such an attitude into their diplomacy and economic relations, although this can be temporarily disguised with a considerable amount of effort and forbearance. Military commentators in the PRC have expressed this perspective to a considerable degree in formulating "legal warfare" as part of the official concept of "Three Warfares."[5] Legal warfare seems to have no corresponding concept among democratic states, where law is seen as the antithesis, or at least a preventive, to warfare. One may make the case that international law has been used as a justification for armed hostilities by Western states, and indeed most court proceedings can be viewed as sort of a verbal "battle" between disputing parties. However, the perspective that the very purpose of international law is to legitimize one-sided declarations or forgo meaningful compromise evokes negotiations conducted by Nazi Germany—the ultimate authoritarian state—as recorded by William Shirer:

> "But this is nothing less than an ultimatum!" Chamberlain exclaimed.
>
> "Nothing of the sort," Hitler shot back. When Chamberlain retorted that the German word *Diktat* applied to it, Hitler answered, "It is not a *Diktat* at all. Look, the document is headed by the word 'Memorandum.'"[6]

This is not to say that authoritarian states cannot be successful in their diplomacy and negotiations, but it implies that prudent decision-makers approach these interactions with a degree of suspicion. This is premised on the understanding that diplomatic relations among democratic governments have a nature and course that are very different from those between democracies and nondemocracies (or authoritarian states) and between authoritarian states and other authoritarian states. Democratic or, more properly, representative or republican governance is described as premised on the "art of the compromise," as well as rule by the majority. There is a psychological tendency, therefore, for modern democratic states to seek compromise and an amiable spirit in diplomatic relations, a spirit that is likely to be reciprocated by like-minded states.

This tendency can be exploited by authoritarian governments, which is why a prelude of diplomatic activity—although without real sincerity—can be an effective anti-access tool. The authoritarian anti-access state wishes to appear interested in negotiation and compromise. Prior to the Pearl Harbor attack, Imperial Japan engaged in a flurry of diplomatic activity, specifically to mask its intent.[7] At the very least, the diplomatic activities can feed the naturally hopeful "but negotiations are continuing" reports appearing in the Western media. The intent is to delay the counter–anti-access power's response.

Deterrent Steps of Counter–Anti-Access

In the deterrence phase, the counter–anti-access state must take steps or initiate activities to ensure that its deterrent posture is credible. A strategically superior power can possess all the components necessary to mount a counter–anti-access campaign yet undermine its attempt at deterrence. Conversely, the strategically superior power can enhance the effectiveness of its deterrence through such actions as follows:

- Make its objectives in the region transparent and distinct.
- Make a clear, simple public case for its objectives. Expect domestic and international criticism but be resolute. In a world awash with contending information, no policy—no matter how logical or altruistic—will not be publicly criticized by those whose interests lie elsewhere.
- Maintain a continuous, meaningful dialogue with its regional allies, partners, and potential partners on the threats to peace in the region, including steps it *might* take in the region if hostilities were to occur.
- Maintain a continuous but firm dialogue with the potential anti-access opponent, including steps it *might* take in the region if hostilities were to occur.
- Maintain support for international agreements concerning the global commons that provide the greatest freedom of action for all states.
- Maintain the continuing physical presence of combat-credible armed forces in the region either through basing agreements with allies or partners, or through the utilization of the global commons, with the realization that such forces will always be at risk and that presence itself is not necessarily sufficient for deterrence.

- Make clear a commitment to support the territorial integrity of any state hosting a regional base for the counter–anti-access forces.
- Encourage the active public discussion and debate of counter–anti-access measures that could be taken in the case of a contingency. Clarify with the potential anti-access opponent that such a public debate is unofficial and the natural consequence of intellectual freedom and expression in democratic societies. Even the criticism and expressed fears of those opposed to counter–anti-access action can enhance deterrence by making the prospect of action appear more credible.
- Embark upon a program to continually upgrade the survivability of regionally based or deployed armed forces based on the evolution of military technology. This should be neither a crash program of improvements (which could signal a negative expectation of current survivability) nor an effort to achieve absolute protection (an impossible goal). Rather, it should appear a gradual, natural process of improvement.
- Conduct out-of-the-region exercises against notional anti-access campaigns. The point of doing such exercises out of the region is to avoid increasing any level of regional tension, but also to reduce the chance of extensive intelligence-gathering by the potential opponent. The objective is to make the potential opponent aware that such exercises occur without the opportunity to examine the operation of systems in detail. This is difficult in an environment of satellite and long-range sensors but can be done with proper planning.
- Conduct routine freedom-of-navigation exercises in areas of possible conflict. Be aware of the vulnerability of the units involved, but discount reactive incidents. Long-term deterrence is best enhanced with a "business as usual" approach rather than a "ramp up in time of tensions" method.
- Make clear that hostile regional actions by the anti-access state will have economic consequences in terms of trade and financial interaction with that state. In a financially globalized world with highly developed trade networks, economic pressure can have particular, yet subtle effects—even against the most isolated regimes.

Measures to enhance the credibility of regional, conventional deterrence will differ based on the circumstances of the region and the potential anti-access scenario. The above listing is certainly not exhaustive, but it is representative of steady, routine actions that counter some of the activities listed in Figure 2. These efforts cannot ensure the success of deterrence. However, academic postmortems have suggested that their absence can trigger a breakdown of deterrence. Evidence is anecdotal but includes Secretary of State Dean Acheson's omission of Korea in describing America's area of interest in Asia in a 1950 speech and Ambassador April Glaspie's apparently ambiguous statement to Saddam Hussein in 1990.[8] Since we have no sure means of measuring deterrence, attending to anecdotal evidence is prudent.

Assessing the Military Anti-Access Campaign

Moving beyond the deterrence phase, one comes to the focus of the majority of studies of anti-access warfare: the military campaign. Through evaluating the sources discussed in Chapter 2, it is evident that there are at least four ways for analyzing or assessing the particulars of an anti-access campaign. Approaches to analysis can consist of the following or a combination of them:

1. sequential elements of the campaign
2. typology of elements of the campaign
3. in terms of the systems and weapons of potential use in the campaign
4. in relationship to unique attributes (such as the fundamental elements discussed in Chapter 1)

Most sources that assess specific anti-access scenarios, such as potential hostilities with the PRC or Iran, utilize the first or third methodologies. Those studies that survey foreign military writings generally use the second methodology and provide a typology of anti-access activities. Some short commentaries focus on the unique attributes of anti-access strategies. This book will also assess scenarios in terms of the fundamental elements previously discussed.

Sequential Elements of an Anti-Access Campaign

A military anti-access campaign would not necessarily use techniques or systems that are different from all multidomain combat. The difference in anti-access warfare lies in the campaign objective rather than the specifics of individual tactical operations. However, it is possible to compile a general sequence of operations that outlines the operational characteristics of a "standard" anti-access campaign. The term *sequential* is used with some flexibility; some of the operations occur simultaneously and most could be repeated until they have effects on the enemy. It is meant to convey that the campaign consists of courses of actions that are dependent on previous actions and are placed on the previous actions to achieve mutual support. This returns us, again, to the image of building a wall.

By comparing the conclusion of the sources cited and historical examples, a list of actions can be compiled in a rough sequential order. In order to be able to compare these actions to other types of campaigns, we will apply the Phasing Model of Joint Publication 5.0 (JP 5.0), *Joint Operational Planning* (11 August 2011). The Phasing Model (illustrated in Figure 3.1) is the Joint Staff's notional illustration of the sequence of activities in a military campaign designed to aid in planning the deployment and employment of forces. In the words of JP 5.0:

> Phasing is a way to view and conduct a complex joint operation in manageable parts. The main purpose of phasing is to integrate and synchronize related activities, thereby enhancing flexibility and unity of effort during execution. Reaching the end state often requires arranging a major operation or campaign in several phases. Phases in a contingency plan are sequential, but during execution there will often be some simultaneous and overlapping execution of the activities within the phases. In a campaign, each phase can represent a single major operation, while in a major operation a phase normally consists of several subordinate operations or a series of related activities.[9]

The Phasing Model consists of six phases. The first phase is the shaping phase (phase 0), which we have earlier termed the deterrence phase, consisting of the primary activities of "prepare" and "prevent." In the anti-access model, the shaping/deterrence phase from the perspective of the anti-access force consists of constructing the anti-access network; continually improving the range and capability of individual weapon and sensor systems; developing doctrine, conducting exercises, and training personnel; and taking diplomatic and economic actions to neutralize regional bases. Diplomatic measures and economic incentives are designed to convince regional neighbors that there is no threat to their security and/or that they should not support an out-of area power.

From the perspective of the power-projection/counter–anti-access power, the shaping/deterrence phase consists of strengthening relationships with regional allies and partners; establishing or maintaining regional bases; routinely deploying forward-presence forces into the region; enforcing international law as concerns the global commons; continually improving the range and capability of weapon and sensor systems; developing doctrine, conducting exercises, and training personnel; and taking diplomatic and economic actions to neutralize hostile activities.

FIGURE 3-1. JP 5.0 Phasing Model

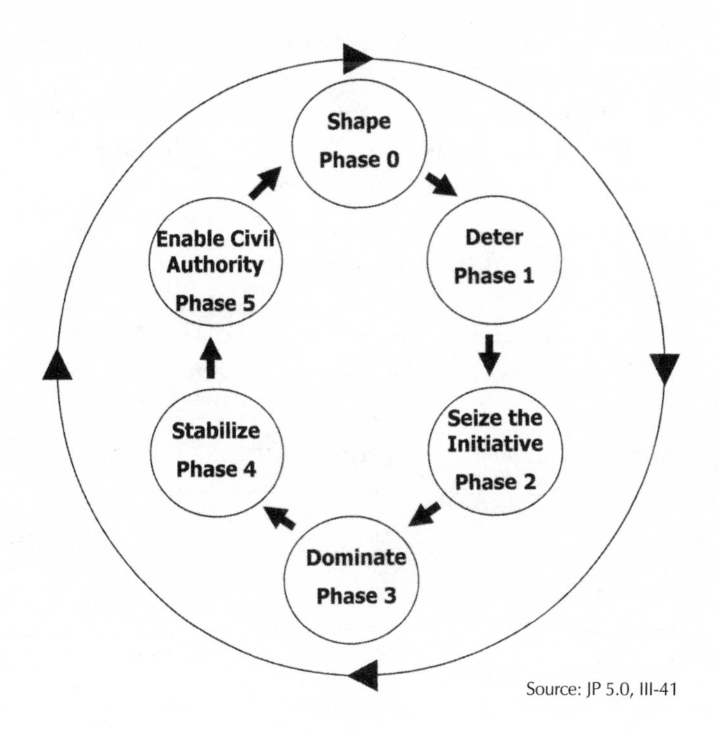

Source: JP 5.0, III-41

JP 5.0 Phasing Model

In the DOD Phasing Model, the next phase is "deter" (phase 1) in which a "crisis is defined." This phase may be appropriate for a Desert Shield/ Desert Storm–type war in which an act of aggression occurs to which the United States and its allies and partners respond after weighing options. But in an anti-access situation in which a strategic premium is placed on preemptive attacks of regional forces, it is best to alter the model to reflect the combining of the shaping and deterrence phases. By the time that the crisis is "defined," it is likely that the initial action of the anti-access strategy is already in motion. In the anti-access model, phase 1 is better conceived as the initiation phase, during which the anti-access forces begin to weaken the regional infrastructure needed for power-projection forces. Starting activities in this initiating phase can include:

- nonattributable/difficult-to-attribute cyberattacks on regional or the potential opponent's infrastructure
- launching and positioning of space-based antisatellite weapons (ASAT)
- use of earth-based ASAT systems to degrade the power-projection/ counter–anti-access force's C^4ISR systems
- initiation of a deception plan for strike forces
- positioning of strike forces at sea, such as submarine barriers, long-range aviation patrols, and sorties of surface ships beyond littoral waters
- establishment of a blockade or maritime exclusion zone
- assault on the regional objective (the proximate cause of military action)
- establishment of air superiority in the region through use of fighter aircraft and ground-based antiair defenses.

These starting activities could be but the prelude to a continuing escalation that includes more intense kinetic activities:

- increasing range and level of cyberattacks
- actual ASAT attacks
- employment of electromagnetic pulse (EMP) weapons
- jamming and deception of other C^4ISR systems
- ballistic missile and aircraft strike attacks on regional bases

- torpedo, cruise, and ballistic missile attacks on forward-deployed naval forces
- reduction of the remaining defenses of the regional objective
- amphibious and airborne assaults on the regional objective
- sabotage or other local disruption of the power-projection/counter–anti-access power's ports of embarkation

Such activities reflect a loss of deterrence, which in phase 1 is more properly conceived as the initiation phase in an anti-access model.

Phase Two as Start of Counter–Anti-Access Campaign

JP 5.0 describes phase 2 as the "seize the objective" phase in which U.S., allied, and partner forces "assure friendly freedom of action" and "access theater infrastructure." In the anti-access environment, this is more properly the counter–anti-access phase in which the power-projection forces struggle to maintain their remaining footprint in the region and bring expeditionary forces into position to attack and penetrate the anti-access defenses. The counter–anti-access phase is characterized by decisive actions on the part of the power-projection/counter–anti-access force in reestablishing combatant forces in the region, including:

- cyberspace counterstrikes
- protection of remaining satellite assets and ASAT attacks on the anti-access state's space systems
- depending on the level of previous destruction to friendly and own C^4ISR systems in the region, use of EMP weapons to eliminate the anti-access state's C^4ISR advantage
- jamming and deception of C^4ISR systems; electronic warfare in the region
- submarine operations in a sea-denial mode within any declared blockade area or maritime exclusion zone
- submarine cruise missile strikes against land-based air defense, C^4ISR networks, and air bases
- tactical strikes against land-based air defenses, C^4ISR networks, air bases, and military leadership targets by remaining regional land-based air forces
- special operations forces (SOF) activities in the region and objective area

- long-range strike by stealth aircraft and bombers against C^4ISR, air bases, missile launchers, ports, and troop formations
- regaining air supremacy in conflict area via carriers and long-range and refueled tactical aviation
- bringing antisubmarine warfare and missile-defense ships into region to provide submarine sanitation (sinking or driving off enemy submarines) and missile defense for counter–anti-access forces and regional air bases; establishment of sea control
- surface ship cruise missile strikes against remaining C^4ISR, air bases, missile launchers, ports, troop formations, surface ships
- reestablishing defenses of regional air bases and air base operations
- amphibious operations in the objective area
- establishment of logistical bases ashore
- airborne operations in the objective area
- movement of maritime prepositioning ships into the region with land combat supplies

Although many of the previous steps are interdependent and require a sequence of supporting operations, others can be performed simultaneously. The elements of the campaign would not be performed in lockstep, but the order displayed previously does correspond to the general flow of counter–anti-access campaign operations.

Phase 3 as Conclusion of Counter–Anti-Access Campaign and Start of Force-on-Force Engagement

In the JP 5.0 Phasing Model, phase 3 is termed the "dominate" phase. The dominate phase consists of "establishing dominant force capabilities" (ashore) and "achieving enemy culmination or joint force commander's favorable conditions for transition." The transition meant is to posthostilities or the "stabilize" phase (phase 4). JP 5.0 describes the dominate phase as focusing "on breaking the enemy's will for organized resistance or, in noncombat situations, control of the operational environment. Success in this phase depends upon overmatching joint force capabilities at the critical time and place, . . . [includes] full employment of joint force capabilities and . . . concludes with decisive operations that drive an adversary to culmination."[10]

Phase 3 assumes that the enemy's anti-access effort has been broken or is fragmentary and sporadic, and the campaign is now of conventional force-on-force battle, leading to the defeat of the enemy's control over the war-precipitating objective.

The stabilize phase (phase 4) includes "establishment of security" and "restoration of services" leading to the "enable civil authority" phase (phase 5) in which the objective area is "transferred to civil authorities" and the power-projection force "redeploys" to its home bases.

In the JP 5.0 model, the counter–anti-access campaign occurs from phase 0 through phase 2, with a transition to decisive operations in phase 3. At this point, the purpose of the counter–anti-access campaign is to enable phase 3 operations through assured access to the region and objective area. Earlier the overall campaign was described as consisting of two (or more) phases; those two phases would be the counter–anti-access campaign and decisive force-on-force operations.

Use of the Phasing Model

Most of the anti-access studies that describe a sequence of counter–anti-access operations do not use the JP 5.0 Phasing Model directly. It has been used here to relate anti-access warfare to overall military campaign planning as directed by joint doctrine. As previously noted, the Phasing Model is a better fit for a Desert Shield/Desert Storm scenario or that of the Iraq War of 2003 than an anti-access environment. In the case of the Iraq War, phase 4, the stabilize phase, and phase 5, the enable civil authority phase, actually became the longest and toughest of the phases.

But the model is useful in eliciting a sequence of operations for the counter–anti-access campaign. Thinking in terms of a sequence causes the analyst to identify the operation requirements that, in turn, can be compared to current or planned operational capabilities. Obviously a campaign can only be successful if the operational force has the capabilities for carrying out these requirements. This is true of both the anti-access and counter–anti-access forces. In a 2002 assessment of then current anti-access efforts, prolific naval strategist Norman Friedman noted that prospective enemies seemed *not* to be investing in the sensors and command-and-control systems that would allow the weapon systems they developed or purchased to be used effectively in an anti-access mode.[11] By 2012 this argument seemed dated—evidence started to build that investments were being made. But Friedman's comment does lead many anti-access studies to focus on identifying the weapons and C^{4}ISR systems that could be used in an anti-access campaign as well as the sequence of operations.

Typology of Elements of the Campaign

The RAND study *Entering the Dragon's Lair* is representative of the typology approach. The primary research methodology is the analysis of PRC writings concerning *local war under high-technology conditions*, the type of conflict that the PLA currently expects to encounter. Local war translates as regional war as used in this book. The primary concern of the PLA is the participation of out-of-the area forces (e.g., the U.S. armed forces), which appear to drive their military investments toward an anti-access approach.

Official PLA doctrine is not publicly transparent. Use of PRC public writings as a surrogate for official PRC doctrine is appropriate because published debates on policy are not permitted without some degree of Chinese Communist Party (CCP) acquiescence—a standard practice in authoritarian states. Published commentary by PRC military officers and academics can be considered, to some degree, to reflect official views or permitted ongoing debates within the PLA.

Entering the Dragon's Lair identifies the following as "elements of Chinese military strategy with potential implications for U.S. theater access":

- attacks on C4ISR systems
- computer networks attacks
- electro-magnetic pulse (EMP) attacks
- attacks on satellites
- attacks on logistics, transportation, and support functions
- attacks on enemy air bases
- blockades
- attacks on sea lanes and ports
- attacks on aircraft carriers
- preventing the use of bases on allied territory[12]

Each of these elements can be considered individually as a tactic of conventional force-on-force engagement. What makes them elements of an anti-access strategy is their combination and their intent. All can be conducted while avoiding a force-on-force engagement. All can be implemented so as to prevent access to the region.

Both RAND and CSBA have examined these elements in detail; their analyses will not be duplicated here. Rather, a few clarifying observations are offered. Although the elements are not specifically placed in sequence, they do follow the overall pattern of methodology 1. Component operations can be simultaneous as well as sequential, and the actual sequence is ultimately dependent on the actions/reactions of the enemy. The typology method avoids tying the anti-access concept to debates on what exactly the anti-access force will do first. The reality is that the anti-access force will seek to do what it can do to prevent the power-projection/counter–anti-access force from entering the regional combat theater. If it can attack satellites without endangering its own capabilities, it will likely do that, but it may not employ EMP weapons for fear of the damage that could be done to its own systems. On the other hand, if its C^4ISR is hardened and its communication systems buried, it might choose EMP as a tool against the power-projection/counter–anti-access force. If the anti-access state does not possess an ocean-going naval force or long-range tactical aviation, it is likely to be unable to conduct attacks on sea lanes and ports. Or it could choose not to do so in order not to antagonize other nations. The typology is a menu of the most likely actions the anti-access state would choose to enact.

Weapons and Systems of Anti-Access

Studies that make specific recommendations for U.S. strategy and defense acquisition programs to counter anti-access warfare usually enumerate the anti-access weaponry that faces power-projection forces now and in the future. The CSBA study *AirSea Battle: A Point of Departure Operational Concept* provides such a listing in its "representative PLA order of battle."[13] As listed in the study, these systems include:

- Kinetic and non-kinetic anti-satellite weapons and supporting space launch and space surveillance infrastructure;
- Sophisticated cyber- and electronic warfare capabilities;
- Long-range ISR systems (airborne; space-based; land-based over-the-horizon radar [OTH-R]);
- Precision-guided conventional land-attack and anti-ship cruise and ballistic missiles numbering in the thousands, that can be launched from multiple air, naval, and mainland-based mobile ground platforms throughout the theater;

- Scores of quiet diesel (and some nuclear) submarines armed with supersonic sea-skimming anti-ship cruise missiles and advanced torpedoes;
- An emerging ballistic missile submarine (SSBM) force;
- Very large inventories (tens of thousands) of advanced sea mines;
- Multi-layered integrated air defense systems (IADS), including a large numerical superiority in modern fighter/attack aircraft, and fixed and mobile surface-to-air missiles numbering in the thousands;
- Comprehensive reconnaissance-strike battle networks covering the air, surface and undersea domains; and
- Hardened and buried closed fiber-optic command and control (C2) networks tying together various systems of the battle network.[14]

The CSBA report provides nominal ranges for aircraft and cruise and ballistic missiles in terms of both distance and possible locations of U.S. forces. The PLA Second Artillery controls DF-4 ballistic missiles, which are likely capable of striking the West Coast of the United States; the D-3A missile with an estimated range of 1,500 nautical miles (nm), which is just short of the distance from the PRC coast to Guam; the DF-21 ASBM with a range in excess of 1,000 nm, which could cover all of Japan including the U.S. naval base at Yokosuka; the DF-15A ballistic missile with a range of 500 nm, which covers Okinawa and southern Japan; the DF-15 and DF-11A ballistic missiles with ranges in excess of 300 nm; and the DF-11 ballistic missile, which is optimized for the 170 nm needed for strikes on Taiwan.[15]

The PLA order of battle also includes JH-7 strike aircraft with an unrefueled range of 800 nm and the H-6 bomber with a range of 1,000 nm. Both planes are capable of carrying the C-803 cruise missile with a range of approximately 150 nm and the larger DH-10 cruise missile with a range in excess of 1,000 nm depending on launch altitude. The DH-10 can also be fired from ground launchers.

This is a formidable array of weapons with ever-increasing ranges and improving guidance systems. The DF-21 ASBM is reputed to be able to conduct reentry maneuvers based on targeting data in order to strike a moving aircraft carrier. As anti-access systems, they have considerable keep-out capabilities. The ballistic missiles aimed at land bases (or cities) can be targeted simply by programmed fixed coordinates, requiring no sensor guidance. Weapons directed at ships or other moving targets require internal sensors or mid-course targeting data or both. But these are systems that the PLA has or is developing, including intelligence, surveillance, and reconnaissance (ISR) satellites; over-the-horizon (OTH) radars; and airborne ISR aircraft.

The advantage of the analysis of anti-access campaigns in terms of specific weapon systems is that it clarifies the degree to which power-projection access is threatened and generates a discussion of possible technological or tactical countermeasures. It also makes the anti-access more real to planners and decision-makers who focus on programs, research and development, and acquisition. From those perspectives, analyzing a potential anti-access campaign in terms of weapon systems is "practical" and avoids the theoretical, definitional, and historical issues that are the heart of this book. More important, it leads directly to recommendations for action by the DOD. For example, if the PRC's DF-21 missile can hit a moving vessel at sea, improvements in ship characteristics, deception programs, counter-ISR programs, or ballistic missile defenses may be called for. Contrariwise, one could argue that U.S. reliance on maritime power is weakened, or that the DOD should avoid investments in certain ship types. In any event, the debate over what course to take can be conducted in relation to programs or tactics rather than strategy.

Driving the focus to a specific weapon system, however, can provide for a confused or incomprehensive debate. An example is the PRC's DF-21 antiship ballistic missile, which has been called a "game changer" in professional literature.[16] The DF-21 may be a fearsome weapon from the perspective of an individual ship, and a saturation attack is a daunting prospect for naval power-projection forces, but a single weapon does not make an anti-access network, no matter how new or symbolic it might be. A focus on specific weapons may miss a total campaign perspective, as expressed by Friedman:

> Yet the battle is not between a missile and a ship, or a submarine and a ship, or a mine and a ship. It is between our fleet and an enemy. The missile or mine or submarine has to come in proximity with the ship and arrange an engagement. In some very important cases, prospective enemies seem not to have appreciated the extent to which other capabilities are needed to make their missiles or mines or torpedoes

> effective against us. Conversely, our own countermeasures may be most effective against elements of enemy force other than the actual weapons.[17]

The weapons-and-systems analytical methodology is most effective when anti-access systems are evaluated as a network rather than individually. Doing so can make the threat of an effective anti-access strategy concrete.

Relationship to Unique Attributes

A fourth way of analyzing anti-access campaigns is by examining what makes them different from previous conceptions of defensive tactics. Of course, that is the primary purpose of this book. But focusing on the unique attributes of anti-access and counter–anti-access campaigns can point to specific defense policies, structures, doctrines, and programs that allow the networks of the campaign to be most effective. Although many observations derived from this methodology may seem quite general, sometimes the general or most obvious observations are the ones most easily missed. A starting point for examining the unique attributes of anti-access are the fundamental elements used through this book. In the next chapter we will add some supplemental factors derived from the other three methods, but at this point we will concentrate on the fundamentals. Because the fundamentals have been discussed in some detail from the perspective of the anti-access state, the discussion later will be primarily from that of the power-projection/counter–anti-access force.

Strategic Superiority

Ironically enough, in terms of grand strategy the fact that a potential opponent would choose to invest its military resources in developing an anti-access strategy is a very good thing. It means that the anti-access state does indeed realize that its objectives could be thwarted and fears the risk involved in a force-on-force engagement. It means that it does see its opponent as a strategically superior power and may act accordingly. In an East Asian scenario, it is impossible to determine whether attempts by the CCP to put a final end to the Chinese civil war and take Taiwan have been deterred by its perception of the strategic superiority of the United States, but such does seem a likely case. Being strategically superior has a deterrent effect of its own. The whole point of an anti-access effort is to mitigate this superiority and reduce the deterrent effect. But this is preferable to the power-projection power than a situation in which the erstwhile anti-access state stops its investment in anti-access systems because it no longer fears defeat in a direct force-on-force engagement.

The great German strategist of land warfare Karl von Clausewitz advises that "the best strategy is always to be very strong, first generally then at the decisive point."[18] Always being very strong would apply well to strategically superior states. Being strong at the decisive point is the challenge for its counter–anti-access force. Defeating an anti-access network, like breaking a wall, requires that strength be applied to the foundations of the network. C^{4}ISI and other sensors are the foundations from which anti-access weapon systems are directed, a thought that lends itself to Freidman's advice that the strategically superior power's strength should be directed "against elements of enemy force other than the actual weapons." But in any event, it is always easier to be strong at the decisive point(s) when one is strong everywhere, because that provides the pool of strength that can be drawn upon.

Retaining strategic superiority does not require one to be superior in every type or aspect of warfare. There are, for example, asymmetrical forms that as Barnett noted earlier may not be acceptable. High-technology nations have the tools to be superior to nonstate actors in the use of terrorism and, like Imperial Japan, might even be able to recruit their own suicide bombers in even greater numbers. But most of them would not choose to do so, and therefore they could be described as inferior in such techniques. Dropping the inherent facetiousness, it should be recognized that "to be very strong" does not require unlimited investment in every military capability that can be possibly conceived. In fact, in the case of counter–anti-access campaigns, the strength that is needed lies as much in the alliance network, supporting logistics, and use of the global commons that are prerequisites to the ability to operate in distant regions at all, than in a particular weapon system or technique.

Strategic superiority, particularly military superiority, is also the product of history as well as effort. The relative global power of both the United States and the Soviet Union from 1945 to 1991 was largely a product of the mutual devastation of the great colonial and European powers during World War II. History thus becomes an element that needs to be taken into consideration in the perception of strategic superiority. An apparently "declining power" that has a history of strategic victory may retain a residual infrastructure and attitude from which strategic superiority can be rebuilt.

In a global security environment in which anti-access strategies appear to become more prevalent, the four key areas of investment that appear most necessary to the projection of power and countering anti-access warfare are the previously mentioned (1) maintenance and defense of regional alliance networks, (2) retention and improvement of the logistical networks and infrastructure that can transport and retain power-projection forces in far regions, and (3) extensive ability to use the global commons—vast and free from other states' sovereignty—as maneuver space and a base from which power-projection forces can operate, along with (4) a military research-and-development system that can routinely develop countermeasures that pace the technologies and techniques of anti-access. A nation needs to maintain its investment in these four prerequisite areas, even at the expense of advanced systems acquisition, if it is to retain a position of strategic superiority.

Once built through history and the prerequisite infrastructure, strategic superiority does not necessarily require a military spending level that threatens to bankrupt the national economy during economic downturns. Strategic superiority—even in the face of anti-access challenges—does not actually require unlimited investment. Acquisition costs are generally related to the technological sophistication of the systems being procured. Precision and stealth are two capabilities that are particularly costly. Both are needed to spearhead the initial penetration of effective anti-access networks. However, every system used does not have to be precise or stealthy. Anti-access warfare is as much about numbers and attrition of systems as technological sophistication. Saturation missile attacks, prodigious naval minefields, and swarming attack platforms are all prominent techniques of anti-access. They can also be prominent techniques of countering anti-access if used creatively and aggressively. For example, ships carrying antiship cruise missiles cannot approach a fleet entering the region if their own port is mined. The point is that countering anti-access can be effective via a mix of the most sophisticated and much less sophisticated weaponry and platforms. For this the budget does not have to break the bank of the strategically superior state.

Strategic superiority is not just a definitional fundamental—it is also a posture from which a counter–anti-access campaign can be built. A nation that can sustain a position of strategic superiority has the capacity to optimize its resources for a successful counter–anti-access campaign.

The Influence of Geography

The geography of the region inevitably shapes both the anti-access and counter–anti-access campaigns. A region such as that including the Persian Gulf with its narrow confines and with but one strait as its maritime entrance presents a different operating environment than the broader dimensions of the western Pacific. All counter–anti-access campaigns have the same general sequence of operations and the same general requirements as to types of military assets. But the relative proportions of these assets needed for success will differ. As comparative examples, the difference in the distances involved in the Normandy invasion of World War II and the penetration of the regional defense of the Japanese empire necessitated or allowed a different mix of assets. With secure regional air bases in Britain, Allied land-based tactical aviation could be used with little requirement for sea-based carrier aviation. In the Pacific, the distances involved in a maritime environment put a premium on aircraft carriers; tactical land-based aviation could not close with the enemy until island bases were captured from the sea. Likewise, an anti-access state whose primary land geography is desert requires or allows a different mix of forces than a mostly mountainous state. The flexibility of multimission platforms seems to be the answer in constructing counter–anti-access forces that can operate in different geographic environments.

Space and cyberspace are domains or dimensions that do not in themselves seem to possess geography. But their access can still be affected by geographic conditions. Communication networks via fiber-optic cables dug into the ground have a different degree of penetrability than wireless systems. Command-and-control nodes can be better protected from attack if they are located in caves or under mountains than in buildings or bunkers in a desert (assuming they have been located). The location of space launch facilities can dictate the direction of launch and the orientation of satellites. Sources have suggested that the resulting narrow windows of the atmosphere-space interface that rocket payloads must pass to achieve orbit could be "mined" by ASAT weapons in the same way as narrow maritime straits. Geosynchronous orbit positions necessary for surveillance of particular earth locations could also be subject to this sort of mining.

To include geographic factors in recommendations for counter–anti-access operational requirements, the best analytical approach seems to be that of the regional scenario. This is the approach that has been taken by CSBA studies.

Predominance of the Maritime Domain

The predominance of the maritime domain has already been belabored. But the fact of predominance does mean that analyzing the ability of a potential anti-access campaign in terms of the level of sea-denial capability is a useful surrogate to more complete measures when other aspects of the campaign are unclear. Conversely, the power-projection/counter–anti-access force's capability to retain sea control is also a useful measure of the overall campaign.

If an anti-access force can achieve regional sea denial, it has succeeded in blocking the flow of additional power-projection forces into the region—eliminating the attack of the follow-on forces, in old NATO parlance. This leaves the destruction of regional land bases as its remaining tactical challenge. As noted, targeting of fixed land bases is an easier problem than targeting maneuvering forces, whether maneuvering on land, on sea, or in the air. Achieving sea denial means that the primary anti-access challenge has been largely overcome. This was the logic of the Soviet Union in the manner in which it invested its resources in naval forces. The ultimate goal in defeating the U.S. Navy was to stop the flow of supplies and reinforcements from North America to NATO Europe, largely the same strategy adopted by the German navy on two previous occasions.

The ability to achieve sea control is a prerequisite for a counter–anti-access campaign in most theaters. Measuring the ability of a counter–anti-access force to achieve sea control in time and breadth is largely a measure of the potential success of the counter–anti-access campaign overall.

Information, Intelligence, and Deception

The criticality of information and intelligence is another point that has been belabored. But few if any of the tactical operations in an anti-access or counter–anti-access campaign can be conducted without very detailed information and intelligence and the resulting targeting information. Possibly the only tactical operation that can be conducted without them are ballistic or cruise missile attacks on land bases of which the targeting coordinates are publicly well known.

Anti-access warfare therefore becomes a struggle to gain information, with space becoming a prime battleground. This is not to say that land-based radars and electronics interception facilities and other sensors are not important, but that they are standard targets of any missile and air campaign. Space assets, on which high-technology nations have become dependent for detailed surveillance and reconnaissance, present a different challenge. There is a paradox: greater reliance on space assets provides more useful information, but greater dependence on space-based information creates greater overall vulnerability in the face of ASAT weapon capabilities. For the United States, reliance on sophisticated spaced-based assets appears to have led to the gradual retirement of land-based "legacy" sensors, certainly a reduction in redundancy and survivability. The United States has also chosen a path of greater reliance on commercial satellite systems, particularly for the transmission and relay of communications and information. At the same time, potential anti-access states, such as the PRC, are increasing the inventory, range, and sophistication of their space-based information collection systems. Perhaps this will make them more dependent on space assets and less likely to employ ASATs and ground-based antisatellite systems.

As we have noted in the discussion of other analytical methods, a struggle to protect or destroy satellites would likely be the opening phase in an anti-access/counter–anti-access confrontation. There is but a limited number of nations that have demonstrated a capability for conducting such operations, although it is possible for all nuclear weapon–capable states to explode an atmospheric or near-space nuclear weapon to create an electromagnetic pulse that could destroy unhardened low-orbit satellites. EMP is, of course, a two-edged sword in that it could also damage the employing force's own electronic systems. But it is a ready tool.

ASAT Systems

Only three nations have demonstrated dedicated ASAT systems: the United States, Russia, and the PRC. Russia, in the guise of the Soviet Union, was the first to demonstrate an ASAT capability utilizing a co-orbital satellite that would gradually approach a target satellite in such a way that its orbit would bring the two into close proximity. The first prototype launch occurred in November 1963, and actual kills (on its own target satellites) were demonstrated in the late 1960s. Total co-orbital ASAT program launches were twenty-three in number, with the system declared operational in February 1973. Testing was resumed in 1976 to develop a higher orbiting satellite weapon for potential use against the U.S. Space Shuttle. In addition to this program, the Soviet Union experimented in the 1970s with ground-based lasers—which are reported to have temporarily blinded several U.S. satellites—as well as directed-energy weapons (which it was unable to perfect) and targeting techniques for use by its space stations. In the 1980s it developed an air-launched ASAT missile to be carried into a high atmospheric firing position by the MiG-31 Foxhound tactical fighter. Following the collapse of the Soviet Union, the Russian ASAT program remain largely dormant until August 2009 when the Russian air force stated that it had resumed the use of the MiG-31 as an ASAT launching platform. It is unclear whether actual tests were attempted, but in May 2010 it was also reported that other ASAT weapons were under development.

The MiG-31–launched weapon was a counterpart to the U.S. military's most practical ASAT weapon, the air-launched ASM-135 ASAT missile carried by modified F-15 tactical aircraft. This program began in 1982, with its sole successful intercept in September 1985. The program was ended in 1988. Previously the United States had experimented with ballistic missile–launched ASAT weapons in the 1960s; however, they appeared to be impractical without the use of a nuclear warhead. In 1962, prior to the Comprehensive Nuclear Test Ban Treaty, the United States tested an atmospheric nuclear warhead to measure EMP effects. The test damaged three satellites, as well as affected communication and power generation across the Pacific region, convincing American decision-makers that employing EMP weapons might be as much a threat to U.S. forces as to an enemy. During the Strategic Defense Initiative, research was conducted on alternative ASAT measures, because the technologies involved paralleled that of potential anti–ballistic missile defenses, but the effort ended with the Cold War. The United States never developed a co-orbital ASAT.

On 21 February 2008, the United States destroyed a malfunctioning spy satellite (whose decaying orbit might have resulted in reentry of its toxic fuel tank) with a modified Standard Missile 3 (SM-3), an anti–ballistic missile weapon fired from an Aegis destroyer in the Pacific. Ostensibly the action was carried out to prevent fatalities or damage on Earth from the fuel tank, although it was widely perceived as a response to an ASAT test by the PRC in the previous year. A motive generally ignored in the media was the destruction of the satellite ISR equipment itself, which could have been exploited by others in the unlikely circumstance it survived reentry. This was the sole ASAT test conducted by the United States following the Cold War and does not actually demonstrate a standard capability, since both the missile and the guiding Aegis radar were substantially modified, and there appear to be currently no other such modified missiles in the U.S. inventory.

The PRC conducted a successful ASAT test against an obsolete weather satellite in January 2007, an action that likely prompted the U.S. SM-3 test. The SC-19 ASAT missile used is a modified ballistic missile with a kinetic-kill warhead fired from a mobile launcher. Demonstration of this capability was internationally viewed as reigniting an "ASAT arms race."

In 2010 Indian officials announced that they possessed the components for ASAT systems and would likely develop such weapons in conjunction with their ongoing ballistic missile defense program. It may be that ASATs have replaced nuclear weapons as public indicators of great power status among the nations of the world.

The purpose in providing such detail on ASATs is to point out their usefulness in any anti-access networks and the vulnerability of counter–anti-access forces that do not have comparable systems or defenses. The criticality of information and intelligence for the counter–anti-access campaign and the reliance on space systems for such information makes the opening phase of such combat potentially decisive. ASATs are frequently mentioned in PRC discussion of their Assassin's Mace anti-access strategy.

Deception is another obvious feature of anti-access/counter–anti-access conflict. If you can't even see the wall, it is hard to take it apart. Deception has been used as a standard element of conventional war and has been demonstrated as an effective tool of anti-access and counter–anti-access efforts in numerous historical examples.

Extraneous Events

Distracting extraneous events have been a regular feature of many campaigns, with anti-access forces capitalizing on their occurrence. Their use will be discussed in greater detail in the historical examples of the following chapters, but some general observations will be made here.

The strategically superior power is the combatant most likely to be affected during the campaign by extraneous events. The obvious reason is that it generally has greater interests in global issues than a regional power. Although his global view was limited in comparison to today, Xerxes is an appropriate example. He had a large empire of conquered peoples to maintain, and suspension of his Greek campaign did not appear to pose a direct threat to his control. His reputation as undefeated would be jeopardized—and it is important for any authoritarian government to appear uncontested. But his campaign could be spun as a victory by his scribes because his army burned Athens, the home of the most troublesome of the Greeks as far as meddling in Ionia was concerned. In contrast, his physical presence in Asia was more important because the Persian Empire was largely "an organization in which there was little incentive to get the job done unless you could cut a fine figure in front of the boss."[19]

Historical examples of successful counter–anti-access campaigns are those in which the power-projection/anti-access force could not be shaken in its resolve. In the total war scenario of World War II, extraneous events just didn't matter. Nazi Germany did hope that the Japanese assault against the Allies and particularly at Pearl Harbor would delay or discourage U.S. involvement in the European war. But having resolved with Britain on a "Europe first" approach and initially undertaking a holding action in the Pacific, the United States did not conform to Hitler's expectations. German and Japanese strategy was also thoroughly uncoordinated. Any event in the rest of the world was deemed secondary to defeating the enemy.

Extraneous events might be fostered or encouraged by the anti-access state, but the most distracting are those that affect the national interests of the strategically superior opponent that have nothing to do with the conflict. Extraneous events that are obviously fostered by the anti-access state can tempt a strategically superior opponent to deal with the source rather than resolve the local problem. Extraneous events, however, become major factors in "wars of choice."

Defeating Anti-Access Warfare

What do the previous analytical methods tell us about the ability to defeat anti-access warfare?

The first is that deterrence is as critical to the strategically superior counter–anti-access power as it is to the anti-access state. In fact, it can be said to be more critical, since the strategically superior power is likely to prefer the antebellum status of international politics. In an East Asian scenario, the United States would prefer that a forcible annexation of Taiwan by the PRC *not* occur. On the other side, the PRC would be most likely to activate its anti-access strategy *in conjunction with* the forcible annexation of Taiwan in the hope that its formidable anti-access capabilities would deter U.S. intervention against its objective. A strategically superior power is most likely to achieve deterrence when it is apparent that it can channel its efforts—as Clausewitz describes it—at the decisive point. This requires defined steps to counter the opponent's anti-access capabilities. Again, the paradox—and defense budget dilemma—is that counter–anti-access forces function best when they don't have to be used.

A second observation is that the necessary requirements for a credible and effective counter–anti-access campaign are defined by a fairly obvious sequence of events or components that constitute anti-access warfare. There is no great mystery about what sort of steps and systems are required. The sources cited provide excellent lists of these components and system requirements. Many of the necessary systems correlate with those required in force-on-force combat, but some do not. To achieve credibility and deterrence, the strategically superior power must be seen as making at least some investments in systems specifically designed for countering anti-access.

Third, the JP 5.0 Phasing Model, which is the basis for joint planning, can be modified to fit the anti-access scenario, but it is not an exact fit. Planning a counter–anti-access campaign more properly requires a sequencing model more in step with the elements proposed by CSBA and others. This might be a composite model, since the counter–anti-access campaign is but the first step in the overall war. The war continues until the opponent surrenders or withdraws from its objective. It should be expected that the AirSea Battle Office will develop and promote an improved model of planning for anti-access warfare. Suitable scenario-specific models already exist, and following chapters will add additional insights to them.

Fourth, the criticality of information and intelligence puts a premium on cyberwarfare but even more so on space warfare. Of the ASAT-capable nations, the United States—the nation most likely to conduct a counter–anti-access campaign—currently seems to have the weakest ASAT program. That is a vulnerability on a par with relying on unshielded digital communication networks. Perhaps there are classified "black" programs that are designed to correct this deficiency.

The final comment for this chapter is that a counter–anti-access campaign is not a military problem isolated from the effect of extrinsic events on the whole of government.

Chapter 4

Three Anti-Access Victories

As concerns international relations—of which war is a part—history is our laboratory. Technology may change, but the motives and character of war do not. War is an attempt to bend an opponent to one's will. We may live in an era of limited objectives and "wars of choice." But this does not invalidate the fact that the objective of war is to compel a change of situation in accordance with one's grand strategy. History is our only true source of experimentation and knowledge of warfare.

Method of Assessment

It is incumbent on us to study anti-access strategies in terms of this laboratory. What follows in this chapter is an assessment of three historical examples of successful anti-access campaigns. Chapter 5 analyzes three unsuccessful anti-access campaigns. The assessment will be conducted from the perspective of the five fundamental elements identified in Chapter 1, but with the addition of four supplemental factors derived from the other analytical methodologies discussed in Chapter 3.

As before, the fundamental elements are:

- The perception of the strategic superiority of the attacking force
- The primacy of geography as the element that most influences time and facilitates attrition of the enemy

- The general predominance of the maritime domain as conflict space
- The criticality of information and intelligence, and—conversely—the decisive effects of operational deception
- The determinative impact of extrinsic events or unrelated events in other regions

The supplementary factors representing areas of primary concern for countering anti-access forces, as expressed in the CSBA and RAND studies and the *JOAC*, are:

a. Strikes against existing regional bases
b. Preemption, including attacks on the ports of embarkation of the power-projection/counter–anti-access force
c. Innovation, including technical, tactical, and organizational innovations
d. Cross-domain synergy

Choice of Campaigns

The cases selected for assessment are meant to be heuristic and illustrative, rather than definitive. Excellent arguments can be made for inclusion of other cases or the exclusion of particular cases selected, on the grounds that there may be better examples. A list of all historical anti-access campaigns does not exist, and compilation of such a list is not within the scope of the present study. Instead, a brief survey of general military histories used in university and war college courses was conducted to identify candidates that appeared to match the anti-access concept. The entry qualification was that the campaign involved a strategically superior power attempting to gain access to a region or area with an anti-access defender trying to exclude outside forces. In some cases, regional access was intended as a prelude to invasion of the defender. In others, it was an attempt to force the anti-access state to relinquish territory or other objectives obtained by force. In yet others, the superior power wanted to operate in the region in order to achieve some other objective of its grand strategy. Cases were considered successful when the anti-access state was able to keep the superior power out of the region, at least until another separate attempt by the superior power, no matter what the final outcome of the overall war. Cases were considered unsuccessful anti-access campaigns when the counter–anti-access forces were able to project power, operate in the region, and achieve their military aims. It is possible that these aims were thwarted or reversed during a posthostilities phase, but that is not considered a disqualifying factor.

Neither the length of time of the campaign nor the size of the forces involved is considered in including or excluding cases. One area of concern, however, is how to define what constitutes a region. Even though regions may not have rigorous delineations, there is some consensus today on how to identify a general region based on geography or culture in the context of the global environment. In previous eras, the globe seemed decidedly smaller because of a lack of knowledge of far areas. In those cases, regions are defined as areas primarily separated by water and with limited direct access by land. This definition has been applied to certain later cases as well.

It might be suggested that such a definition promotes a tautology since it reinforces the argument that the predominance of the maritime domain is a fundamental aspect of the anti-access situation. My response is that I could not find a better method of identifying historical regions in a way that parallels what the sources in Chapter 2 identify as modern anti-access environments. Therefore, I will let others argue regions and appropriate cases, and simply proceed.

The cases included are not meant to be detailed and are not based on research in original sources. Some background information is included, and the outlines of the campaigns are sketched. The information is gathered from both scholarly and popular sources. But the extent of the information provided is only sufficient enough to support a discussion of the anti-access strategies behind the campaigns. Readers are urged to consult other sources in order to gain a fuller understanding of the history. The cases should in fact be considered vignettes.

The three cases chosen as successful anti-access campaigns—anti-access victories if you will—are (1) England's defeat of the Spanish Armada, (2) Ottoman Turkey's defeat of Allied efforts to force the Turkish Straits in World War I, and (3) the Battle of Britain in World War II.

Case 1: The English Defeat of the Spanish Armada

The defeat of the Spanish Armada and of the planned invasion of the British Isles by troops positioned in the Spanish Netherlands in 1588 is considered a significant event in world history that ensured the survival of the Protestant Reformation. It was the most decisive operation in a nearly thirty-year cold war between Queen Elizabeth I of England and King Phillip II of Spain. The cold war turned to open hostilities following the execution in 1587 of Mary Stuart, former queen of Scotland and claimant to the English throne. The conflict exhibits the primary characteristics of an anti-access campaign on the part of the English forces, the result being defeat of the Spanish invasion attempt and a subsequent increase in English global influence.

The campaign is extensively documented.[1] There are numerous scholarly analyses of the Armada campaign utilizing extensive contemporary records from English and Spanish participants.[2] Official reports were published and publicly distributed almost immediately following the defeat of the Armada. A number of the operational commanders wrote or dictated public memoirs—notably Lord High Admiral Charles Howard and Sir Francis Drake—that were published during their lifetimes. More recent, scholarly interest revived in the late 1940s and 1950s, partly inspired by apparent parallels with World War II's Battle of Britain, and peaked in the 1980s with the four hundredth anniversary of the event and as result of extensive underwater archeology conducted on Spanish wrecks off the coasts of Scotland and Ireland. This has resulted in revised interpretations concerning technical aspects of the engagement, but the strategies and tactics of the campaign are not in dispute.

Initiation and Progress of the Campaign

Planning for an amphibious assault and conquest of England can be traced at least back to an August 1583 proposal from Spanish admiral the Marquis of Santa Cruz, who had led the conquest of the Portuguese Empire. Phillip II requested and received from Santa Cruz a detailed plan for "the Enterprise of England" that called for 150 heavily armed great ships, 360 more ships for transport, supply, and support missions, and a landing force of 64,000 soldiers—a unified and self-contained expeditionary package that Spain could not then afford. Phillip also requested a plan from his top land commander, the Duke of Parma, then conducting a war against Dutch rebels in the Spanish-owned Netherlands—a rebellion partly subsidized by Elizabeth. Parma proposed the movement of his existing (or, rather, reinforced) army of thirty thousand infantrymen and four thousand cavalrymen via barges across the narrowest part of the English Channel near the Flanders coast. With the crossing conducted in a single night, this would presumably not require a massive naval force. Parma, however, stipulated that complete surprise and avoiding the English navy were essential elements for success, a stipulation that Phillip responded to with the annotation "Hardly possible!" on his copy of the plan.[3]

Phillip developed a less costly, hybrid plan that combined both proposals into one for a naval armada that would defeat the English navy in the Channel and convey Parma's force across from the Flemish port of Dunkirk (now French). Phillip's final approval for the execution of the plan would depend on events, but his operational commanders were aware of their responsibilities in 1586, at least six months prior to Stuart's execution.[4] Through spies, code breaking, and diplomatic sources, the English government was also aware of the existence of such a plan, and this awareness along with the realization that Spain would rely on the support of English Catholics—who might rally to Stuart as rightful queen—was likely the major factor in prompting Elizabeth to acquiesce to her advisers' demand for her death.

The English decision-makers recognized that Spain was militarily and strategically superior. The English army was vastly smaller than that of the Spanish Empire and was designed to be employed in foreign wars overseas. More important, English home defenses on land were locally organized and uncoordinated, primarily relying on poorly trained levies. England's military strength lay in its navy, although that too was poorly organized and supplied. Throughout her reign, Elizabeth insisted on minimal expenditures, and a number of the wealthy, noble captains used their own money to pay their crews and provide supplies. She did, however, have a ready supply of brave sea captains with great experience in naval combat, navigation, exploration, and—in some cases—piracy. These captains had already developed a reputation of technical, tactical, and organizational innovation. They viewed their ships as floating batteries, with their sailors to serve the artillery and conduct occasional raids ashore, unlike the Spanish view of naval combat as land combat at sea, decided by grappling, boarding, and close engagement by embarked infantry.

With knowledge of the planned Armada, Queen Elizabeth was willing to back a "private" expedition in 1585 and 1586 led by Sir Francis Drake against Spanish trade in the West Indies, the primary source of Spanish wealth. Two royal ships participated, with the rest of the fleet financed by backers. Although the expedition was not a financial success, Drake seized ships and destroyed ports and towns in the West Indies and the Spanish Main, delaying the sailing of the Spanish treasure fleet by many months. By mid-1586, King Phillip was having severe financial problems, which affected his military operations under Parma against the Dutch, who had reinvigorated their rebellion.

With Phillip continuing his plans for the Armada, Elizabeth sent an official fleet under Drake in 1587 to conduct a preemptive strike on the Spanish force in its main port of embarkation, Cadiz, and "to impeach the joining together of the King of Spain's fleet out of their several ports, [and] to keep victuals from them."[5] Ironically enough, Elizabeth changed her mind concerning Drake's entry into Spanish ports, but Drake had (perhaps deliberately) put to sea before receiving the revised order. Intent on destroying all the moored and anchored ships and, if possible, port facilities, Drake embarked about a thousand troops in his fleet. Following a rendezvous off Lisbon, Drake forced the defended harbor of Cadiz directly, burning or capturing from thirty to thirty-four ships, and destroying a vast quantity of stores being collected for the Armada. Historians credit Drake's "singeing of the Spanish beard" with delaying the Armada by almost a year.

Aware that a sailing date for the Armada was soon to be set, Elizabeth began mobilizing her fleet in January 1588. The Marquis of Santa Cruz died in February, allowing Phillip to appoint a more pliable but much less experienced commander for the Armada, the Duke of Medina Sidonia. On 28–30 May the Armada began to set sail, this time out of Lisbon (Portugal having been captured in 1581 by Phillip), but storm damage forced it to put in at the Spanish port of Corunna for repairs. Elizabeth was again determined to destroy the Spanish fleet before it could arrive in the English Channel and sent Lord Admiral

Howard and Drake to attack it in Corunna, but contrary winds forced them to abandon the expedition and return to England.

The Armada sailed from Corunna on 12 July bound for the Channel and the planned linkup with Parma for the escort of his troops from Dunkirk to England. It was sighted off the coast of Devon on 29 July, and the English fleet, delayed by a headwind, was out to meet her by the next day. The English fought three major battles from 31 July to 4 August in the Channel against the Armada, thwarting it from landing an advanced party on the English coast, damaging ships, and demoralizing its captains and crews. The Spanish had expected to grapple and board the smaller English ships and overwhelm them with soldiers. Instead, the English engaged them with artillery at long range and then, finding their own ships more maneuverable than those of the Spanish, fired with greater effect at shorter range while still being able to avoid being boarded. Several Spanish ships were abandoned by their crews and captured, although the overall Armada maintained order.

Both fleets anchored on 6 August off Calais (now France), where Medina Sidonia found Parma inexplicably but thoroughly unready, having not yet embarked his troops and supplies into barges. It became apparent that it would take Parma two weeks to complete that phase. On 7 August the English attacked the Armada with fire ships, and the Spanish ships cut their cables and put to sea. The English fleet then inflicted a decisive defeat on the Armada off Gravelines (near Dunkirk), with nearly half of the Spanish ships too heavily damaged to fight again. With his fleet disorganized, demoralized, and running short of supplies, his captains insubordinate, and a wind forcing him out of the narrowest parts of the Channel and into the North Sea, Medina Sidonia elected to return to Spain by sailing around the north of the British Isles and into the Atlantic. Many ships were wrecked on the coasts of Scotland and Ireland, with only 60 of the 130 ships that had initially set sail in the Armada returning to Spain.

Elizabeth tried to follow up this anti-access success by sending another expedition under Drake, this time to Lisbon to reinstall the king of Portugal, but he was unsuccessful. Phillip built new ships for another invasion attempt a decade later, but they never reached the Channel. In the meanwhile, the obvious defeat of Spanish sea power resulted in a swarming of English, Dutch, and French privateers, as well as pirates, into the West Indies to harass and seize the Spanish trade for many years.

Elizabeth also renewed her support for the Dutch struggle against Spanish troops. Among the many poems and hymns written to honor the English victory, one by William Verheyden of Leiden (Netherlands) was titled *Oration on the Fleet of the Spanish Xerxes*.[6]

Assessing the English Anti-Access Campaign: Fundamentals

The fundamental elements are defining characteristics of the English anti-Armada campaign. England recognized that Spain was *strategically superior* (1) in global affairs and militarily superior on land. Spanish colonies in America provided her vast wealth; England's colonies were not a source of wealth but actually a deficit. English armies had not proven overly successful against Spanish arms in the Netherlands. The English homeland defense force was disorganized and poorly trained. Elizabeth's councilors knew they had to keep Spanish troops off English soil.

The *geographic characteristic* (2) of the English Channel was and had previously been the primary planning element in English wars against continental rivals. Considerable attention had played on Parma's force throughout his operations in the Netherlands, which was a scant distance across the Channel from Dover. But to get his forces across required at least temporary sea control by the Spanish fleet. The *maritime domain* (3) was therefore the *predominant medium* in which the English could contest Spanish entry onto their territory.

English knowledge of the planning of Phillip's "Enterprise of England" gave it an advantage in terms of the *criticality of information and intelligence* (4). Although the actual date of the invasion was unknown, its general outlines were well known and could be planned for in advance. Both sides used spies and reporting from diplomatic missions to provide local intelligence. Phillip was provided assistance from some of the English Catholics, particularly those who surrounded Mary Stuart, but Elizabeth's councilors were aware of these machinations, and other English Catholic nobles stayed loyal to the queen. Neither side seemed to have achieved results from deception.

Extrinsic events (5) played a role in at least two ways. First, the Spanish forces under Parma were engaged in a continuing war against the Dutch, a war that the English supported and subsidized. Although historians condemn the unready Parma for his negligence (as did Phillip II), his focus was naturally on his own central front rather than defeating England, the glory of which he would have to share with Medina Sidonia. Elizabeth supported the Dutch not only for their Protestantism, but because they diverted attention and resources away from her territory.

More important were events completely outside of Europe. Drake and others raided the Spanish colonies in the New World, the riches of which Spain had become economically dependent. Although the English did not cut off the trade, it was convoyed in fear of their depravations—in a fashion similar to Mahan's depiction of a fugitive on the sea. A popular rumor in Spain was that Drake had a "magic mirror" in which he could discern the location of Spanish ships. It was not that the English controlled the sea at that point in world history, but that the Spanish could not deny English use of the sea in areas that were critical to Spain. English operations in the Indies were more than a distraction.

Assessing the English Anti-Access Campaign: Supplementary Factors

Analyzing a historical case using the supplemental factors is meant to relate the case to current aspects of anti-access strategies that are of most concern to strategists. The supplemental factors are operational in nature, whereas the fundamental elements are characteristics of the environment or on the strategic planning level.

The act of identifying *strikes* (A) against existing regional bases in the Spanish Armada campaign bumps against the difficulty of defining "region." Are the British Isles to be considered a region separate from continental Europe? Is Northern Europe, which would include the Netherlands as well as the British Isles, to be considered a region separate from Southern or Mediterranean Europe, which would include Spain? Even today, Northern and Southern Europe are often referred to as separate regions, particularly in disputes over economic and Eurozone issues.

Based on the technology of the day and the distances involved, a good argument can be made for the Northern Europe/Southern Europe division of regions. The distance between Dunkirk and Dover is 47 miles (75 kilometers). The distance from Lisbon to Dover is 1,302 miles (2,095 kilometers). With the range of artillery in the hundreds of meters, these are dramatically different distances. It is valid to argue that Parma's bases in France, Belgium, and the Netherlands were *regional* bases—that is, in the same region as England. Spanish and Portuguese ports would be considered "out-of-area."

There are no reports of substantial English raids against regional bases. One likely cause is that "conjunct operations"—the contemporary term for army-navy amphibious operations—were exceedingly difficult given the technology and military organization of the time. Parma also possessed the advantage in number of troops and had a formidable reputation as a successful field commander.

The English did conduct a *preemptive strike* (B) against the Spanish fleet. They also intended to conduct additional attacks against the Spanish ports of embarkation, although they were thwarted by environmental factors. Drake's preemptive attack delayed the sailing of the Armada for almost a year, during which Elizabeth could strengthen her defenses.

English use of *innovative tactics* (C) was definitely decisive in the engagements at sea. Naval combat was transitioning from the close-range cannonades, ramming, and boarding more appropriate for the Mediterranean than the Atlantic. Rowed galleys were still the primary war vessels of the Mediterranean and remained in the Spanish inventory (although only four were part of the Armada). The English, with a heritage of long-range Atlantic seafaring, had abandoned the galley almost after the Viking age. English captains refused to grapple and board, since the Spanish had superiority in musketeers and pikemen. Instead, they maneuvered and stayed at gunnery range. It is a misconception that English gun technology was superior to that of the Spanish; in fact, the Spanish guns were equivalent. It was the tactics and training that were not.

In translating *cross-domain synergy* (D) into the Elizabethan Age, one could say that it consisted of land forces (army) working in close coordination with maritime forces (navy). But if so, one must admit that such operations were uniformly abysmal. The English could mount successful raids in the Americas but on a small scale and not in the face of determined resistance at the water's edge.

Of the supplemental factors, English efforts at preemption and innovative tactics were the decisive operational elements of the anti-access campaign.

An additional observation concerning the Armada is that the Spanish commander, Medina Sidonia, was neither soldier nor sailor, but merely a trusted relative of King Phillip. In contrast, the originally assigned commander, Santa Cruz, had been a victor of the famed Battle of Lepanto against the Turkish fleet. Medina Sidonia tried to refuse command of the Armada. Poor command and control and limited operational competence were also critical factors in the Spanish defeat.

Case 2: Ottoman Defense of the Turkish Straits

The attempt of the Allied Powers—particularly Great Britain—to capture and control the Turkish Straits in 1915 and 1916 was intended to break the stalemate of the trench warfare of the western front by providing a warm-water route to resupply Russia and reinvigorate the eastern front against the Central Powers. Although it created an enduring legend of Australian courage, it has not generated as many studies as other aspects of World War I and is often treated in general histories as a misguided and muddled mistake.[7] But even a severe critic of the then lord of the admiralty Winston Churchill, who was the most vocal proponent, concedes that Churchill's "proposal to outflank the Central Powers—Germany, Austria-Hungary and their allies, including Turkey—by attacking the latter at the weakest link in a front deadlocked from Belgium to the Balkans—is now widely accepted as the boldest strategic concept of the First World War."[8]

The attack in the Turkish Straits (which included both the Dardanelles and the Bosporus, with the Sea of Marmara in between) was intended to have strategic effects. Although Ottoman Turkey had become one of the Central Powers, it posed little in the way of a tactical threat to Allied operations. Germany may have wished for the Turks to pose a threat to the SLOCs and land routes from the Mediterranean to British India, but it never had more than a nominal capacity to do so. But its very existence ensured that there was no linkage from the Mediterranean to a failing Russia. Without passage to the Black Sea, Allied troops could not be brought from the western front to the eastern front, and Russian or other Allied forces could not be brought to the west. Balkan states on the side of the Allies could not be supported. Perhaps even more important, supplies of grain that were vital to the war effort could no longer be shipped from southern Russia. This, combined with the U-boat campaign, was seen as an element in the German effort to starve the British homeland.

Historians divide the struggle for the Turkish Straits into two phases: the naval Battle of the Dardanelles and the combined operation (amphibious assault) at the Gallipoli Peninsula. Each ended as a failure of British planning and determination. Only after the end of World War I did it become known from the leaders of Turkish forces how close both efforts had come to success owing to Turkey's logistical weaknesses.

Initiation and Progress of the Campaign

The Ottoman Turkish Empire activated its secret treaty with Imperial Germany in September and October 1914 by closing the Turkish Straits to Russian merchant ships and conducting an attack on the Russian Black Sea coast. Russia and its British and French allies, engaged with Turkey's allies of Germany and Austria-Hungary, declared war on Turkey. The Turkish navy was commanded by the German admiral Wilhelm Souchon, with the Turkish army commanded (initially in part, then completely) by German general Otto Liman von Sanders. Both Souchon and Liman von Sanders recognized the weakness of Turkish fortifications at the Dardanelles and took immediate efforts to strengthen them. The British Imperial War Council concluded that capturing the Turkish Straits was the principal means of outflanking the Central Powers, but that it required a combined army-navy assault to take the forts defending the Dardanelles by both land and sea. At its narrowest, the Dardanelles is only 0.75miles (1.2 kilometers) wide but with a length of 38 miles (61 kilometers). The shoreline, particularly on the European side, is rugged with many suitable locations for well-positioned and defended artillery.

Having concluded that a combined operation was necessary for success, the British secretary of state for war, Field Marshal Lord Kitchener, insisted that no British or imperial troops were available for the operation, which he at the time viewed as a distraction from the western front. Allied troops were also not then available. First Lord of the Admiralty Winston Churchill directed the Royal Navy to plan the forcing of the Straits by naval forces alone. By February 1915 some army units were identified as potentially available, but by that time the navy-only effort had commenced.

The Allied navy in the naval battle of the Dardanelles was substantial: eight Royal Navy and four French battleships, along with at least four light cruisers, twenty-two destroyers, seven submarines, thirty-five minesweeping trawlers, and a seaplane carrier. Seaplanes were used for reconnaissance and targeting of the Turkish forts. The British had taken the island of Lemnos, commanding the entrances to the Dardanelles, and garrisoned it with Royal Marines for use as a logistical base. Gradually more troops were being sent to the region to exploit any success that the naval force might achieve.

The Turkish fleet made few efforts to attack the Allied navy at sea but had placed naval minefields at the entrances to the Straits, a routine anti-access tactic. Initially, the Allied navy had success against the outermost forts and commenced minesweeping operations. This turned into a difficult prospect, since Turkish artillery began to focus on defending the minefields rather than dueling the ships. Royal Navy rear admiral Rosslyn Wemyss, the commander of the base at Lemnos, noted that in his view, "the battleships could not force the straits until the minefield had been cleared—the minefield could not be cleared until the concealed guns that defended [it] were destroyed—they could not be destroyed until the peninsula was in our hands, hence we should have to seize it with the army."[9]

After eighteen days of inconclusive naval bombardments (with intervening days of bad weather), Lord Kitchener changed his views. British and dominion land forces, spearheaded by Australian and New Zealand Army Corps (ANZAC) troops, began arriving at Lemnos. By 18 March the Royal Navy thought it had all the minefields identified—a fateful mistake—and undertook a massive, final, navy-only attack on the Dardanelles forts, operating destroyers and battleships up the Straits. But the battleships began to strike an unknown mine line, with the first, the French *Bouvet*, capsizing within minutes. Two British battleships sank. Three more Allied battleships were badly damaged by gunfire. The Allied navy elected to withdraw, ironically enough at the time most of the Turkish guns were out of ammunition and with all minefields finally identified.[10] The perception of a disastrous failure permeated the British Cabinet. The combined operations phase began with many questions about the wisdom of the entire campaign.

Liman von Sanders opted for a forward but not (initially) fixed defense, with a mobile reserve. Five rows of trench lines were dug at key locations, the Turkish forces to be moved to wherever the Allied forces landed. The mobile reserve was commanded by a particularly brave and energetic officer who led from the front, Mustafa Kemal—later to become Attaturk, founder of the modern Turkish state. On 25 April 1915, the Allied forces landed at seven beaches nearly simultaneously. One, on the Asian side of the Dardanelles, was a feint that succeeded in tying up a Turkish division and was the focus of Liman von Sander's attention until he discerned its true purpose. Landing on the beach in force without substantial loss, the Allied forces could not push to their objectives on the high ground against rough terrain and Turkish forces that began to move into their trench lines. Part of the problem was caution on the part of British officers with experience in the trench stalemate on the western front, who were reluctant to advance against any opposition even if light.

The effect was a re-creation of the western front, with some opposing trenches within eight yards and the same grenades thrown back and forth between. The Turks could not push the Allies into the sea; the Allies could not push the Turks off the heights. In the bloody battles that ensued to take the no-man's-land between positions, casualty rates reached 50 percent.

By 7 June the British war council already perceived that the attack was a disaster, forming a Dardanelles Committee to determine whether to withdraw. Even Churchill suggested negotiating a separate peace with Turkey in exchange for removal of the Allied forces. The Turkish government was nearly bankrupt, and even there the sense of fatalism was strengthening. Churchill's suggestion was turned down by the council owing to the likely effect on Russian morale. Russia was determined to defeat the Turks in the Caucasus and achieve territorial gains. The British and Australian press was highly critical of the campaign. There were a number of changes of officers involved, including the British general in command.

By December the war council decided to evacuate all the 83,000 troops. This was accomplished by 8 January 1916, with only three additional soldiers being wounded during the withdrawal. Churchill had been ousted from the Admiralty and the Cabinet in May 1915, but fortunately such a setback did not deter him from continuing his political career. A Parliamentary Dardanelles Commission was established to assess blame. In Turkey a ruthless persecution of Armenians and any other non-Turkish peoples deemed as traitorous ensued. With the Russian Revolution beginning in 1917, the British campaign against Ottoman Turkey was continued in Arabia and Mesopotamia. Turkey surrendered with the other Central Powers in October 1918, and the Ottoman Empire collapsed into chaos.

Assessing the Ottoman Defense of the Turkish Straits: Fundamentals

Even though Britain and its allies were facing a multifront war, its *strategic superiority* (1) vis-à-vis Ottoman Turkey was never questioned. Turkey received at least eight hundred German officer advisers, but in terms of resources it was on its own.

It was *geographic conditions* (2) that prompted Allied interest in Turkey at all. In this the Straits were both a blessing and a curse. They were optimal for an anti-access defense, but they were the prize that has always attracted invaders, including the original Turkish tribes themselves. There were few other reasons for the Allies to focus on Ottoman Turkey, a country that they had defended against Russia just forty years before. Now Russia was the ally, Turkey the enemy. Turkey conducted engagements against Russia early in the war but generally without success. Nevertheless, the Turkish Straits permitted a formidable defense against Mediterranean forces seeking to capture the capital of Constantinople. The land terrain was rugged and difficult to pass, making the sea route the critical factor. But ships had to pass within close range of shore batteries, and naval mines could close the Straits.

The *maritime domain or medium* (3) was a predominance factor in that the struggle was over the use of straits, with control of land only important as it permitted or prevented use of the straits. Combat at the Gallipoli Peninsula was on land, but the Allies could only be sustained by sea. Without the Royal Navy possessing absolute sea control, the operation was nearly impossible. Even with absolute sea control, the operation was unsuccessful. It must be admitted, however, that maritime predominance was not a factor of the Turkish anti-access defense itself, so the argument for predominance is weaker in this particular case.

Information and intelligence (4) is critical in all war. But in this case its lack was an advantage to the defender. The British did not have accurate information on the disposition of Turkish forces before the assault, even as the Turks could not be sure of the intended landing areas. Liman von Sanders based his strategy around this mutual lack of information.

Turkey also made good use of cover for protection of their gun emplacements along the Straits. Using seaplanes to spot minefields gave the Allies an unfounded belief that their knowledge was complete. The seaplanes were much less successful in spotting hidden gun emplacements, and the weather made their use over land difficult. The Ottoman Empire did not possess an air force, so the British forces could have capitalized on airborne reconnaissance and attack but did not bring such assets to the theater in sufficient numbers. Finally, the Allies did not have intelligence concerning the state of Turkish logistics or the ammunition remaining for the Turkish guns.

Extrinsic events (5) were significant factors in British decision-making. Kitchener initially refused to take any troops away from the western front or other duties. This prompted the initial navy-only attack. By the time of the amphibious operation, the Allies had lost any strategic surprise they may have had. At the same time, the naval wars in the North Sea and against U-boats were the Royal Navy's main focus and the Dardanelles expedition largely a peripheral concern—of great value if successful but not critical if lost. Support for Russia may have been important, but eyes were on the western front. In contrast, the anti-access campaign was the Ottoman Empire's main focus even if there were military operations occurring in Arabia and the future Iraq.

Assessing the Ottoman Defense of the Turkish Straits: Supplemental Factors

Without an effective high-seas navy or an air force, the Ottoman Turkish forces had no means of *attacking regional bases* (A), such as the island of Lemnos. British forces were also staged in Alexandria, Egypt, once an Ottoman province but now independent under firm British influence. On land, Turkish forces could potentially make strikes against Salonica (Thessaloniki) in Greece, which had a pro-Allied government and was available for Allied troops supporting anti-Turkey Balkan states. But although Greek control over territory formerly part of the Ottoman conquest may have rankled Turkish nationalists, it was not seen as a direct threat at that time. The Turks also did not have the capability to conduct a *preemptive strike* (B).

To a certain degree, the Royal Navy did try to use *innovative tactics* (C), such as submarine penetration of the Straits and operations in the

Sea of Marmara and Black Sea. But this was a very difficult passage owing to an undercurrent to seaward, and few submarines were able to perform this feat. Those that did engaged in commando-style operations such as dynamiting Turkish train tracks, as well as torpedoing military logistical craft intended to support the Dardanelles defenses. But they were few in number and did not achieve significant operational effects. The combined amphibious assault was also the largest that had ever been conducted up to that time and logistically went superbly. Some troops were landed on the wrong beaches, but supplies moved forward despite Turkish artillery attempting to hit transports and small craft. Lessons learned from Gallipoli were among those incorporated into amphibious innovations prior to World War II.

Cross-domain synergy (D), or the lack of it, was a critical factor in impeding the counter–anti-access force. British navy and army operations were tactically coordinated but operationally uncoordinated. Royal Navy gunnery was well directed against the forts in the Dardanelles but was not used as effectively for close support of the troops ashore. Admittedly, the Allied and Turkish trenches were very close, and gunnery could hurt blue forces. But the Allied troops could potentially have fallen back and allowed major bombardment. The lack of effective use of air assets is also conspicuous. On the other side, with the exception of naval mining—inherently an independent operation—Turkish forces could only operate in one domain, making synergy a moot point.

Case 3: British Victory in the Battle of Britain and Operation Sea Lion

The Battle of Britain has been intensively studied and is considered a turning point in World War II.[11] Prior to launching the air attacks on the British Isles, Nazi Germany had been successful in all its blitzkrieg campaigns, although the Kriegsmarine (German navy) sustained crippling losses against the Royal Navy and Royal Norwegian Navy during the battle for Norway. The Battle of Britain itself was completely an air campaign, although the intent of the Luftwaffe (German air force) attack was to gain air superiority over the Channel and Britain in order to force a negotiated peace or cover a cross-Channel amphibious assault by the German army. The battle was conducted in the summer and autumn of 1940. The result was a defeat of the Luftwaffe in what Prime Minister Winston Churchill called Britain's "finest hour," and Hitler's cancellation of Operation Sea Lion, the planned amphibious crossing.

Initiation and Progress of the Campaign

On 10 May 1940 Nazi Germany invaded France, and Winston Churchill became prime minister of the United Kingdom. Over the objections of Royal Air Force (RAF) Fighter Command, which was short of pilots and knew it was depleting its home defenses, Churchill sent fighter squadrons to fight in France. There they sustained heavy losses while the French defense collapsed. British and French soldiers were evacuated from Dunkirk, and France surrendered on 22 June. Exploring the possibility of invading the Soviet Union, Hitler believed that the British would sue for peace in the wake of the Fascist conquest of most of the European continent. But Churchill and the new Cabinet were determined to fight on, and Hitler discussed plans for a cross-Channel invasion with Grand Admiral Erich Raeder, the Kriegsmarine commander in chief. Raeder insisted that the Royal Navy could only be neutralized if Germany achieved air superiority over Britain and then concentrated Luftwaffe assets on the maritime battle in the Channel. Hitler apparently agreed with this assessment but directed Raeder on 16 July to draft a plan for the amphibious assault. A major prerequisite of this "Directive No. 16: On the Preparation of a Landing Operation against England" was that RAF attrition had to be sufficient to ensure it could not conduct sustained air attacks against the German crossing.

On 1 August the German high command issued the plan for conducting the assault, code named Operation Sea Lion, with mid-September as the planned time frame. The prerequisite task of air superiority was turned over to Field Marshal Hermann Göring, chief of the Luftwaffe, who expressed great confidence of success. Despite the secret nature of the plan, he sent it to air fleet commanders via coded radio messages, which the British intercepted and decrypted. German planning for the air battle and crossing of the Channel was therefore not a surprise.

Initial Luftwaffe estimates of the length of time it would take to destroy RAF Fighter Command were ridiculously low, representing the overconfidence from recent successes. After high command study, the

Luftwaffe was assigned the five weeks of 8 August to 15 September to achieve temporary air superiority over Britain so that the crossing could proceed. The Luftwaffe plan started with destroying coastal RAF bases and then proceeding against inland airfields, particularly those protecting London. Following that, a massive air offensive would strike all military installations and military industries, particularly aircraft construction. If air superiority could not be achieved, the option of conducting a terror bombing campaign against the British population was examined, but Hitler—still hoping for a negotiated surrender—forbade it at that time.

Historians have suggested that the initial Luftwaffe campaign plan was muddled in that disputes arose among its commanders concerning whether the RAF could best be destroyed in the air or on the ground. In the opening phase, the Luftwaffe and the RAF engaged in the *Kanalkampft* over the English Channel, with the Germans attacking British supply convoys moving coal for industrial production. With heavy damage to the ships, convoys through the Channel were suspended and coal transported slower by rail. This may have prompted a degree of German optimism concerning their efforts, but the kill ratio of aircraft favored the RAF from the start.

The German objective in their main assault was to trigger the British air defenses and destroy the responding RAF fighters in direct combat, the assumption being that the superior Luftwaffe fighters would achieve high kill ratios. This was supplemented by bomber attacks on airbases and, eventually, any possible target. In preparation for their massive sorties, the Luftwaffe did commence with attacks on radar stations. However, these appeared unsuccessful—with the British often able to restore operations in about six hours after damage. The Germans suspended attacks on the radar network, ostensibly because of the lack of results but more likely because the Germans did not understand the effectiveness and importance of radar to the British defense network.

German efforts could not defeat the networked British air defense system, which combined intelligence, detection, centralized command and control, antiaircraft units, and positive fighter direction with the new technology of radar. Only on a few days of the campaign did the Luftwaffe destroy more RAF planes than it lost. Aircraft losses became unsustainable for the Germans, and British defense industries were not crippled.

By mid-August 1940, the Germans shifted their raids toward the industrial infrastructure of cities. Sources claim that the first night raid on London was the result of mistaken attacks in violation of Hitler's instruction not to bomb the city except at his personal direction. But then the RAF responded by launching attacks on industrial targets in Berlin, and Hitler's prohibitions were removed. In a series of raids involving four hundred bombers and six hundred fighters targeted on London, the Blitz became a war of cities.

Contradicting the assumption of air power theorists in the 1930s that the bombing of cities would paralyze society and destroy their citizens' will to fight, British popular determination strengthened. By mid-September Hitler began to question whether to end the air attacks in order to conserve aircraft and trained pilots. Recognizing that air superiority could not be achieved, Operation Sea Lion was postponed, though not cancelled. Official German records list the air campaign as continuing to May 1941, but the date generally cited in Allied histories as its conclusion was 31 October 1940. On 22 June 1941, Germany commenced Operation Barbarossa against the Soviet Union, and then resources for a cross-Channel invasion were no longer available.

Assessing the Battle of Britain as an Anti-Access Campaign: Fundamentals

Selecting the Battle of Britain as a case reinvigorates the question of whether the English Channel delineates regions. For the Armada, it was pointed out that the Spanish fleet needed to voyage over 1,000 miles (1,600 kilometers) to enter its objective area. German aircraft in the Battle of Britain flew from captured airfields in France and the Low Countries. Yet many of these sorties were conducted at extreme operating ranges for fighter aircraft, and aerial refueling was of the future. In this context, they were operating at long distance. Certainly the Channel—under the effective control of the Royal Navy—presented as difficult a water barrier as it had nearly four hundred years before. But there is another analytical reason to include the Battle of Britain as an anti-access victory worthy of study: it was the only such case that was fought almost exclusively in the air domain. The air domain is a particular focus to many current studies of potential counter–anti-access campaigns.

The fundamental elements are present in this case. Although Britain had the resources of its dominions and possessions to call upon, Europe was essentially under the control of Nazi Germany. The British army had been pushed off the continent. France, its most powerful ally, had surrendered. The United States had not entered the war and had a strong isolationist sentiment. The Royal Navy remained the most powerful in the world, but its use against a continental enemy seemed limited. The Luftwaffe possessed more aircraft and had proven itself in combat. Even members of the British Cabinet had argued for a negotiated peace. At this point in the war, Nazi Germany seemed *strategically superior* (1).

Again, the *geographic characteristic* (2) of the English Channel remained Britain's greatest defensive asset. Weather patterns were also an important consideration in the air war and can be considered one of the geographic characteristics. As long as the Royal Navy controlled the sea approaches to Britain, an invasion force could not get across. Even without air superiority, it was likely that the Royal Navy would sacrifice ships to tear up the crossing German platforms, something it was willing to do in the early stages of the war in the Pacific. German amphibious craft were poor, and the Germans' experience of amphibious operations was limited. They considered parachute and glider assaults unlikely to succeed. With a stronger navy, Britain had the Channel as its wall.

This meant that although the combat operations of the Battle of Britain were almost exclusively in the air (and in responding to damage inflicted on ground targets), the maritime domain remained the *predominant medium* (3). If England were to be invaded, the enemy had to cross sea against a defending naval force that the Germans judged to be too strong.

The *criticality of information and intelligence* (4) was revealed in the 1970s when the records of the extensive and amazingly successful British signals intercept and cryptographic effort were declassified. The famed cryptology station at Bletchley Park had broken the German Enigma code. The British high command knew much about German planning for the attack, as well as the composition of aircraft formations and commanders' intent, and had early warning of some—but not all—of the targets. The British leaders chose to sacrifice some targets in order to preserve the secret of Ultra, the information obtained from Enigma intercepts. In contrast, German signals intelligence, reconnaissance, and battle damage assessment were weak. Although the early research on radar had been done in Germany, it was Britain that invested much in developing it as a practical sensor. It is doubtful that the Luftwaffe ever grasped the extent of British reliance on radar or its true effectiveness. The Germans did make use of tactical deception, but their strategic deception efforts were stripped away by their reliance on radio communication and confidence in their codes. Acquisition and disseminating of information was a great advantage for the anti-access forces.

Extrinsic events (5) or, rather, an extrinsic event was a hinge upon which the German effort would swing. Hitler was more committed to an attack against the Soviet Union than the invasion of Britain. He also believed, even against British tenacity, that Britain would eventually agree to a negotiated peace. The U-boat fleet, although not yet employed to the extent it would be, appeared to be an alternative means of forcing a British surrender. Racial ideology (against the Slavs to the east) also played a factor in this inclination. If it had concentrated its resources on the cross-Channel operation, it is possible that the German army could have gotten significant forces across, although at heavy loss. But its leaders did not have the opportunity or desire to effect such a concentration. Britain was an enemy but not the primary one. The Germans fixed their gaze to the east.

Assessing the Battle of Britain as an Anti-Access Campaign: Supplemental Factors

Perceiving its resources to be too limited to risk, the Royal Air Force did not conduct extensive *strikes* (A) against German regional bases. But these bases were actually in captured territory that had already been fought over during the opening phase of the war, a different situation than that anticipated in modern anti-access scenarios. In 1937 there had been "a revolution in British air policy" that substituted fighter-centric air defense for the previous bomber-heavy posture of the RAF.[12] This repudiation of "the bomber will always get through" ideology of early air forces was partly motivated by government economy: building fighters was cheaper. But it did lead to a force structure that was most suited to an air defense of the islands and Channel. In that sense, the British sacrificed the ability to strike against regional bases for increased defense in depth.

Preemption (B) was not a strategy of the Allies in War World II. The initial Axis attacks caught them off guard, even in the face of evidence that they were likely. This was largely owing to disbelief in the thirst for war and stated objectives of the Axis powers.

Innovation (C) was a major factor in both the anti-access and counter–anti-access campaigns. It does not seem, however, to have given an overwhelming advantage to either side. Both nations had contemporary high-technology infrastructures and possessed leading scientific communities capable of developing and engineering new weapon systems (relatively) rapidly. Innovation may not have matched innovation directly; as has been discussed previously, the commitment to and effective use of radar appears to have been a major British advantage. But German aircraft were of the highest quality, with combat-proven characteristics, and German scientists were making strides in rocketry. Missiles would eventually be used to strike at the British Isles, although they were not ready for a significant role in the Battle of Britain. The British did create a major tactical innovation in their formation and resourcing of the RAF Fighter Command. Also, the Germans seemed not to initially grasp the inherent vulnerability of unescorted bombers, although there was no lack of tactical innovation by German forces in World War II. In the Battle of Britain, they were primarily outfought.

It is in the Battle of Britain that one glimpses the effects of the modern construct of *cross-domain synergy* (D). The *JOAC* describes this as resulting from close coordination between forces operating in multiple domains. It is tough to argue that the RAF and the Royal Navy were closely coordinated except at the highest level of strategic decision-making. But Britain had the capability of operating with mutual support in both the air and maritime domains or mediums, and Germany did not. The Kriegsmarine had been built as a coastal defense and sea-denial/commerce-raiding force, not a sea-control navy, and it was largely ineffectual in the face of the Royal Navy. There was little spirit of cooperation in the Nazi war councils, and once the task of achieving air superiority was handed off to the Luftwaffe, it is hard to see much effort made by the Kriegsmarine (if at all) to participate in the overall campaign. Much of the planning for Operation Sea Lion belonged to the army, which clashed with the navy over the size and scope of a cross-Channel attack. Although the German war machine used air and land forces in close cooperation in blitzkrieg, air-land-sea domain coordination was poor. British forces came closest to achieving cross-domain synergy in the sense that they maintained forces capable of operating with substantial combat power across the spectrums. This included their effective signals intelligence and cryptology efforts, which can be seen as roughly equivalent to the cyberforces of today.

Lessons of the Counter–Anti-access Campaigns

Studying the anti-access victories of the past gives clues to the problems and prospects of counter–anti-access forces in the present day. Each of the power-projection/counter–anti-access states in the examined cases was strategically superior on a global basis to its opponent. Yet each had difficulty in translating this strategic superiority to local superiority at the point of attack. Each also faced a determined opponent that was willing to risk almost all its military strength in the anti-access effort. In contrast, the strategically superior powers had engagements and interests in a variety of other locations, most of which were a great deal of distance from the location of the anti-access region. Spain had light forces scattered in the Americas and other locations, but the majority of its forces were engaged in the Low Countries. It expected to be able to use some of those forces against England, but this was something to which the local commander was not committed. Great Britain of World War I had a global empire to protect (admittedly the self-governing dominions were doing much of their own regional defense) and had the majority of its forces on the western front or on the North Sea. Nazi Germany in World War II was ruling conquered countries with shocked but resistive populations, and its focus was on the east. The Germans also assumed that Britain could no longer pose a threat to the continent, and so their military strength was positioned or flowed elsewhere. Britain still had its global empire (although soon to be attacked by Imperial Japan), but the most technologically productive part of the world was now under Nazi control or neutral.

Geographic conditions were part of the difficulty facing the attackers, yet none of the counter–anti-access forces were ignorant of these conditions. Each had presumed that their forces could overcome geographic constraints, and in the latter two cases had trained to do so.

But the anti-access forces all optimized their forces to the geographic conditions. Elizabeth poured her resources into a navy that could defend her realm in the Channel or strike Spanish assets elsewhere. Her father had spent money on fortifications, but apparently she did not view such expenditures as particularly efficient. Turkey, which was often in political chaos, had limited opportunities to spend its resources elsewhere and little inclination to do so, although its outlying provinces might be garrisoned. But it viewed Russia as its primary threat and so clustered its best forces around Constantinople. From 1937 onward, Britain optimized the RAF for home defense. Even with Churchill's urging, the commander of Fighter Command, Air Chief Marshal Sir Hugh Dowding, was reluctant to send his planes to defend France in the early stage of the war. He thought that attrition there would leave the British Isles less defensible when the inevitable assault came. His force was optimized for an anti-access campaign while the Royal Navy guarded the Channel. The German air force excelled in close air support of army units in a combined arms campaign, but the Battle of Britain was not a combined arms campaign for them.

Maritime forces were not always the forces primarily engaged, but the maritime environment was always a factor impeding the counter–anti-access force. Spain had the mighty Armada but could not achieve sea control. Even if Parma's troops had been ready, the probability of their crossing the Channel without very substantial loss was low, particularly after the Armada was forced out of Gravelines. Even with command of the sea, the British could not push through the Turkish Straits, take Constantinople, and thereby force Ottoman Turkey out of the war. Nazi Germany's navy simply could not take on the Royal Navy directly. Many historians think Operation Sea Lion never had a chance.

The fog of war is a natural limit to information and intelligence, but the Battle of Britain stands out. In the role as anti-access power, Britain had a source of information the Germans did not suspect—their own coded messages. If a counter–anti-access force allows its own information, intelligence, and plans to be used against it, it is hard to see how its campaign can succeed.

In all cases, extrinsic events elsewhere played significant, perhaps decisive roles in the outcomes. With an eye to the resources needed for other commitments, the will of the counter–anti-access forces to achieve victory in the face of determined opponents wavered. In the case of the Armada, the will may have never been present; Medina Sidonia commanded the force reluctantly and with the premonition of impending disaster. The British Cabinet—governing a democracy—simply could not withstand the popular outcries that had developed concerning the stalemate in Turkey. It never made a good public argument for that campaign. In any case—or shall we say every case—the anti-access defender was simply more determined than the counter–anti-access power.

Strikes against regional bases used by the counter–anti-access force were not a factor in these examples, largely because of a lack of forces and the limitations of the technologies involved. In the case of the Battle of Britain, the RAF could have struck the German air bases, but the decision-makers saw that as a dissipation of the defensive strength they needed. All the counter–anti-access forces were able to use regional bases, a situation not expected in modern anti-access scenarios—and with good reason.

The English fleet used preemption to good effect against Spain, but it was simply not practical in the other cases. Ottoman Turkey did not have the forces to preempt. In essence, the Nazis had already used preemption in striking Poland, and the blitzkrieg was successful elsewhere. The British had been forced off the continent; there was nothing to preempt.

Innovation in tactics, organization, and technology occurred on many levels in the three cases, but in no case was it a significant factor for the counter–anti-access force. Instead, innovation in tactics and organization was used by the English against the Armada. British innovations and technological advantage did not translate into defeat of the Ottomans. The German advisers and commanders did bring organizational innovations to the Ottoman army, but they simply brought Turkish forces to European standards rather than exploited novel ideas. The Nazis never recognized how dependent the RAF was on the relatively new innovation of radar.

None of the counter–anti-access forces was able to achieve cross-domain synergy. Even when domains were limited to two, true army-navy coordination never was achieved on the operational level. Parma just refused to cooperate. The Ottoman and Nazi navies were ineffectual.

There was little the German army could do while sitting on the other side of the Channel.

How do these facts translate into the modern scenarios of anti-access vs. counter–anti-access? In a very real sense, they are warnings to the counter–anti-access forces. If they are to break down the walls, they must proceed with forces tailored to the task, great determination, and a willingness to commit resources and sustain casualties in what is likely to turn into an attrition battle. If the resources and, more important, willingness are not there and extrinsic events occur, they will not be successful.

Chapter 5

Three Anti-Access Defeats

The previous chapter detailed three historical campaigns in which anti-access forces were able to prevent power-projection forces from entering their regions. For the Central Powers, the region was the Black Sea. In each case the power-projection/counter–anti-access forces suffered from poor planning, logistics, and/or cross-domain coordination, and in every case was more concerned about possible extrinsic events than intensifying their own efforts in the face of determined opposition.

This chapter analyzes three cases in which the counter–anti-access forces overcame determined anti-access efforts. Two of these campaigns are very well known, so their background will not be described in any detail. Rather, their anti-access elements are identified along with the methods by which they were overcome. Although the third case occurred more recently than the others, specific details may have faded from collective memory, and a bit more background information is included. The three cases are (1) Nazi Germany's defense of Fortress Europe (Festung Europa), (2) the Pacific War strategy of Imperial Japan, and (3) the Argentine attempt to retain the Islas Malvinas/Falkland Islands following its successful invasion.[1]

Case 1: Nazi Germany's Defense of Fortress Europe

The general history of World War II is well known to readers. This treatment will focus on Nazi Germany's anti-access efforts to prevent the Allies from returning to the continent.

In early spring 1942, the Americans' Joint Staff planners had already begun work to invade the continent with one concentrated thrust through France. Having agreed to a "Germany first" strategy for the global war, the United States proposed Operation Sledgehammer, a limited-area amphibious assault in France in 1943 with a cross-continent push from the beachhead. The British Chiefs of Staff Committee preferred a preliminary course of "closing and tightening the ring" around Germany through a series of amphibious landings or raids around the European periphery prior a major direct assault.[2] One reason is that it felt that Germany was too strong for a direct assault despite the fact that Hitler was committed to an eastern front war against the Soviet Union. The result was a compromise in which an invasion of Vichy France–held North Africa was conducted in 1942 and an assault on Italy started in Sicily in July 1943.

Ironically enough, it was probable that the Allies would have found a landing in Normandy to be an easier prospect in 1942 or 1943 than on the actual D-Day in 1944.[3] The Germans had not yet built the Atlantic wall of Fortress Europe, and as efficient as the German army normally was, regional defense was not focused. However, even if Allied casualties would have been relatively light in the amphibious operation, whether the drive across France could have succeeded at that stage in the war raises some doubts. But what focused German attention on the western wall was the Allied amphibious raid on Dieppe in Nazi-occupied France on 19 August 1942. Dieppe was a port assault rather than a landing on an isolated beach and a military disaster for the Canadian forces that made up the majority of the raiding force. From the experience, the British planners learned that ports were not appropriate targets of direct assault but could best be taken from the rear following a landing at a rural beachhead. The Germans took the opposite lesson and began to turn ports into fortresses, which was fortunate for the Allies because it meant that resources that could have increased defense at rural beachheads were utilized for port defense.[4]

Even so, by 1944 German defenses on the French coast had become more formidable. In late 1942 Hitler had expressed his concerns that the French, Belgian, and Dutch coasts were vulnerable, and he ordered the transfer of the Gross Deutschland Division from the eastern front to France. However, a very controversial book claims that "from the documents available it is clear that Hitler did not fear any serious risk of "losing western Europe" until 1943.[5] Whenever he may have actually feared a serious risk, by 1942 the Germans were building major fortifications that would eventually consist of thousands of permanent coastal emplacements with weapons ranging up to 16-inch naval guns.[6] More important perhaps, in November 1943 Hitler personally ordered Field Marshal Erwin Rommel from northern Italy to inspect the Atlantic Wall from Denmark through France.

Although there were physical vulnerabilities at various locations owing to poorly sited coastal defenses, Rommel's most serious concern was "a lack of doctrine for the defense against a landing."[7] In fact, he took exception to the primary German headquarters directive calling for counterattacks in event of a landing by mobile armored forces held in reserve at a centralized location. Rommel's preference was for local counterattacks supporting a direct defense at the water's edge. As an experienced tank commander, he was doubtful about the movement of armored forces across long distances in the face of likely Allied air superiority and combined arms support.[8] His view was that a counterattacking armored force needed to be kept close to the landing beaches rather than at a farther, centralized location.

Rommel's views were opposed by other German commanders, so Hitler ruled on a compromise that split panzer forces between commanders and doctrines. In order to make up for a shortage of manpower, much of which was committed to the Pas-de-Calais area of northern France, long considered the likely point of Allied attack, Rommel directed the construction of an "unprecedented" barrier. Only a portion was in place by D-Day on 6 June 1944, but sections were formidable. Rommel envisioned the placement of 200 million land mines to turn the French coastline into a vast minefield. About 4.5 million were actually in place by D-Day, still a significant number.[9] Potential glider landing zones in the rear area were defended by "Rommel's asparagus," upright poles topped with explosives. At the beach, successive waves of anti–landing craft obstacles, naval mines, antitank walls, ditches, gun emplacements, and other obstacles were constructed. Innovative systems such as shore-based torpedo tubes and remote-controlled explosive minitanks were adapted. The German navy operated fast-attack craft, conventional submarines, and minisubmarines off the coast. In fact, fast-attack craft sank two Allied Tank Landing Ships (LSTs) conducting a training exercise off England on 28 April.[10]

German defense against the actual landing did not go as Hitler and his generals expected. As has been frequently retold, the immense Allied deception effort that created notional armies in Scotland and southern England, including the "First U.S. Army Group" (FUSAG) ostensibly commanded by Lt. Gen. George S. Patton, fixed Hitler to the preconceived notion that the attack would come in the Pas-de-Calais area and in Norway. The components of the deception included false radio signals, dummy tanks and equipment, and the Double-Cross System in which all the German agents in Britain had been turned and provided false, or true but misleading, information. As with Xerxes, and as is true with most all successful deception, the effort was designed to reinforce what the Germans already preferred to believe—the knowledge of which had been obtained by the Allies through the Ultra operation.

Because it was hardly possible to conceal the fact that the Allies were going to attempt a landing somewhere in Northern Europe, the objective was to convince the German anti-access network that the actual Operation Overlord landings in Normandy were but feints and that the main amphibious assault would be elsewhere. The use of superior weather forecasting, impressively maintained radio silence, command of the narrow sea, and innovative landing equipment such as the portable Mulberry harbors, enabled one of military history's greatest surprises. Noted historian Stephen Ambrose claims that "not one German submarine, not one small boat, not one airplane, not one radar set, not one German anywhere detected this movement." He quotes the deputy head of operations for the German supreme headquarters, General Walter Warlimont, as admitting that German leaders "had not the slightest idea that the decisive event of the war was upon them."[11]

Rommel himself was on leave during the initial landing. Fortress Europe—the leading edge of the Nazi anti-access effort—was effectively breeched through deception, a point that even the Allies' supreme commander, Gen. Dwight Eisenhower, later acknowledged. Much hard force-on-force fighting, and Allied casualties (to say nothing of Russian casualties on the eastern front), had to occur to force the way to Berlin and the final surrender. But the Nazi anti-access campaign was defeated, and power-projection forces were operating ashore with air superiority. All knew that German defeat was now a matter of time.

Assessing the Nazi Campaign: Fundamentals

The fact that the Germans were then operating in an anti-access mode (even if that term did not then exist) was clear to all participants, including the German themselves.

By late 1943, it was widely recognized by German leaders that the Allies were now *strategically superior* (1). A sense of defeatism started to pervade the professional military officers, which eventually led (following the Normandy landing) to tacit support for Claus von Stauffenberg's assassination attempt on Hitler and a failed military coup. Facing an imminent two-front attack on the German heartland, with the staggering reversal on the eastern front and an impending surrender of Italy, the only thing the Germans could hope for was to prevent greater Allied access to the continent until some sort of negotiated peace in the west occurred. The probability of the western Allies agreeing to a negotiated peace was low, but that and Germany's development of some decisively destructive weapons were the only hopes left. The focus on strengthening Fortress Europe and Rommel's desperate effort to force a decision on the beaches was an acknowledgement that if Allied armies were on the ground in Northern Europe in force, they could be delayed but not stopped.

Once again, the English Channel was the decisive *geographic characteristic* (2) on which the German anti-access effort was centered. France, the Low Countries, and western Germany were suitable terrain for tank warfare, albeit not to the same extent as the east. The Channel was the only full barrier. Thus, the maritime domain was the *predominant medium* (3) for the breeching of the anti-access network, even if the hardest fighting was to occur ashore and in the air over land. Air superiority was a requirement, but it always is in a modern maritime operation. The short distance across the Channel from Britain allowed the Allies to rely on land-based tactical aviation and bombers, rather than requiring carrier groups. (In the Atlantic, carrier aviation and naval long-range patrol aircraft were primarily used in the war against the U-boats and in the landings in North Africa and the Mediterranean.) The short distance combined with the military technology of the time meant that reliance on the Channel as the prime geographic feature on which to base the anti-access defense was a very thin reed. But it remained the primary reed.

The *criticality of intelligence and information* (4) and the use of deception as elements that determined success or failure of the anti-access and counter–anti-access campaigns go without saying. The Germans were deceived, their own intelligence network was broken, and through signals intelligence (SIGINT) and Ultra the Allies knew the German command's intentions almost at the same time its own field commanders knew them.[12] The sophistication of the Allied SIGINT and cryptologic efforts is indicated by a continuing series of overt minor actions (such as emplacing additional sea minefields) specifically designed to cause local German forces to report them to headquarters. In this way, the Allies could—knowing what was being reported—determine the actual meaning of the coded signals themselves.[13] Except for "dumb luck," a successful German anti-access effort was not possible in an environment in which the Allies had information supremacy.

Extrinsic events (5) were simply not going to shake Allied determination to breach the wall and prosecute the war until Germany's unconditional surrender. The western Allies agreed on a "Germany first" strategy, and nothing in the Pacific War was going to sway them away from this—at least not in 1944. The Soviet Union had no intention of fighting Japan, and Japan had reciprocated in this tacit agreement. In fact, extrinsic events went completely against Germany. Its efforts to create intrigues in South America amounted to nothing, and the alliance with Japan ended up being worthless. Hitler's absurd and unnecessary declaration of war on the United States following the Japanese attack on Pearl Harbor justified America's involvement in Europe. The Allies portrayed the Nazis as the epitome of evil—which they were—and events elsewhere in the world were not going to stop the crusade.

An assessment in terms of the fundamentals outlines the German effort of 1943 and 1944 in the west as an anti-access campaign and also explains why it failed.

Assessing the Nazi Campaign: Supplementary Factors

The Battle of Britain can be considered the Nazi attempt to conduct *strikes against regional bases* (A). Britain was always the obvious staging area for all the forces for a cross-Channel landing, earning the title of "unsinkable aircraft carrier." But by the time of the actual landing, the Germans had lost control of the air (and never had control of the sea). Even if they had correct intelligence as to the disposition of Allied amphibious forces, the Germans no longer had the capability to conduct sustained attacks. The counter–anti-access force had eliminated this possibility.

World War II started as a war of *preemption* (B), a tool the Germans perfected as blitzkrieg. Preemptive attacks allowed it to defeat an alliance of Britain, France, and Poland that—on paper—was strategically superior. But by the time Nazi strategy evolved by necessity into an anti-access effort, preemption was no longer a possibility.

The German scientific-technical community was the source of considerable *technological innovation* (C), and the German military was a considerable source of tactical and administrative innovation throughout the war. In fashioning Fortress Europe, it had deployed such sidewise technologies as land-based torpedo tubes and remote-controlled explosive minitanks. But the significant technological innovations came too late in the war to mitigate Allied strategic superiority. It is a sobering thought to consider what might have happened if Hitler had delayed the start of the war to 1945 or later to perfect and build an extensive inventory of ballistic (V-2) and cruise (V-1) missiles and jet aircraft. At the same time, the Allies utilized technological innovation, the atomic bomb being the most dramatic—along with considerable tactical and administrative innovations. Both the Allied and German militaries were learning organizations that attempted to adapt rapidly to change wherever possible. The German's greatest drag on tactical innovation in the later stages of the war was Hitler himself, who had taken personal control of German military strategy.

The *cross-domain synergy* (D) between air (attack aircraft) and ground forces that the Germans displayed during blitzkrieg had greatly dissipated by the time Germany was forced on the defensive. With the exception of the U-boat force, German maritime capabilities were limited, and coordination of naval forces with the other services was poor. Luftwaffe chief Göring had championed independent roles for air forces, such as achieving an "inevitable victory" in the Battle of Britain. Although he was capable of close cooperation with ground commanders, it was not necessarily his focus. At the point of shifting to an anti-access strategy, the potential for cross-domain synergy—such as attacks on regional bases and preemption—had passed.

In contrast, the Allied counter–anti-access effort achieved considerable cross-domain synergy by the time of the Normandy landings. Part of this was the result of naval and air superiority (even supremacy), but it also reflected the close interservice cooperation that had been learned through trial and error, as well as a realization that the independent strategic bombing campaign on German industries (and cities) was not achieving the expected results.

Cross-domain synergy was also achieved in the Allied SIGNIT and cryptologic efforts, with operational forces being utilized to provoke coded German radio reporting on known subjects. The ability of the Allies to keep their cryptologic successes from German intelligence—even despite a series of leaks in American newspapers—still remains astounding.

Cross-domain synergy, innovation, control of the prime geographic barrier, maritime supremacy, and strategic superiority itself were all attributes that the Allied counter–anti-access force used to enter Fortress Europe. But the Allies' greatest operational advantage was the information and intelligence obtained through cryptology and SIGINT, and the ability to combine the information advantage with deception. The breaching of the Atlantic wall was truly one of history's most successful military deception operations. Strategically the most significant attribute was a will for victory that was reflected in adoption of a Germany-first policy. There were simply no extrinsic events that could possibly dissuade the Allies from their course.

Case 2: The Pacific War Strategy of Imperial Japan

The history of the Pacific War part of World War II is also well known. After having achieved conquests in China, Indochina, Indonesia, and outlying islands, Imperial Japanese forces shifted to an anti-access strategy in order to ensure that outside powers could not reverse the gains.

From a planner's perspective, the Japanese Empire was now self-sufficient in resources and could over time increase its national power to rival that of the former colonial empires. British and Dutch forces in the region were minor threats because the Netherlands had been conquered by Germany and Britain was fighting for its homeland. Australia, a British self-governing dominion, could attempt a defense in the southern waters, but did not have the manpower to push north of New Guinea and could be reserved as a later invasion target. Japan had come to a tacit agreement with the Soviet Union—soon to be fighting for its existence against Germany—and the Soviets did not declare war on Japan until the Pacific War was almost over. India was a potential conquest or "liberated area," depending on one's perspective. The only power left to challenge Japan's absolute dominance of the Asia-Pacific region was the United States.

For a brief surrogate for details on Japanese anti-access planning—which was based on an awareness of the U.S. forces' long-developed War Plan Orange—we will examine the logic of Naval Marshal General (Fleet Admiral) Isoroku Yamamoto, the commander in chief of the Imperial Japanese Combined Fleet. History is full of ironies; Yamamoto was not only the most senior Japanese officer reluctant to go to war against the United States, but he was one of the first leaders to fall victim to the tactic of targeted assassination when his aircraft was ambushed by American warplanes with the knowledge that he was aboard.[14]

Yamamoto was an authentic expert on the United States, having studied at Harvard and served two tours of duty as a naval attaché in Washington. He was convinced that the country's industrial power and innovative nature would make up for any military weakness it might have and that Imperial Japan could not defeat the United States in a long war of attrition. In short, he viewed the United States as having the potential to remain strategically superior to even an expanded Japanese Empire. These beliefs caused him to oppose the Japanese invasion of Manchuria and China (even then a U.S. trading partner) and the Tripartite Pact with Nazi Germany and Fascist Italy. He was subsequently the target of assassination threats by fanatic nationalist officers.

However, once the army-controlled government decided on war with the United States, Yamamoto devised a strategy to prevent it from returning to the western Pacific by a preemptive strike against the most powerful military assets that could be brought into the region in support of War Plan Orange. This was combined with simultaneous strikes on the Philippines, an American possession, as well as Malaya and the Dutch Indies. Following these successful actions, Yamamoto's plan for the navy was to convert to an anti-access posture that can accurately be described as "to consolidate the outer defensive perimeter of their newly conquered empire and to counter an anticipated Allied offensive."[15] Both the Japanese and War Plan Orange envisioned a decisive naval battle within Micronesia or closer to Japanese home waters. Supporting this decisive battle would be an "island-hopping" (or, in the case of Japan, an "island-defending") campaign across the central Pacific and Micronesia, which provided considerable opportunities to attrite U.S. forces before the decisive battle. Preempting the U.S. fleet at Pearl Harbor and fighting a war of attrition throughout the Pacific were the main features of the anti-access efforts. Considering the war in China to be its primary task and assessing the Americans as lacking in martial spirit, the Japanese army acquiesced to this anti-access posture. Years after the war, a Japanese army planner stated that despite public skepticism by navy leaders, the army simply "did not worry about the Pacific Ocean. We were confident of the Navy's ability to hold back the United States fleet. . . . We trusted them; they didn't come out and warn us of their limitations."[16]

While a tremendous surprise, the attack at Pearl Harbor was a failure—it did not achieve the strategic aims required for Japan's successful prosecution of the war. It did not knock all the chess pieces off the board; U.S. aircraft carriers were not in port at the time of the attack, and key logistical facilities such as fuel tanks were not destroyed. More important, it *increased* rather than decreased America's will to fight, a result that Yamamoto apparently recognized once he realized that Japanese diplomats had failed to make a declaration of war immediately prior to the actual attack. The United States was now fighting for revenge as well as strategic position. Earlier, he had predicted that if the shock of Pearl Harbor did not convince the United States to withdraw from the conflict, "it would not be enough that we take Guam and the Philippines, nor even Hawaii and San Francisco. To make victory certain, we would have to march into Washington and dictate the terms of peace in the White House."[17] Since such was always an absolute impossibility, the true implication was that the remaining sum of Japanese strategy would be but an attempt to neutralize the strength of superior U.S. forces until time, attrition, and/or extrinsic events shook American determination—an unlikely prospect. Yamamoto also accurately predicted, "In the first six to twelve months of a war with the United States and Great Britain I will run wild and win victory upon victory. But then, if the war continues after that, I have no expectation of success."[18]

The naval Battle of Midway in June 1942 is often cited as "the battle that doomed Japan" because Japan lost much of the carrier air strength with which to center both its offensive and anti-access efforts. Although structured as an offensive operation to capture Midway—the outermost island of the Hawaiian chain and located almost in the middle of the Pacific—as well as several of the Aleutian Islands, the operation was actually a trap to lure the American fleet (and most importantly, its aircraft carriers) within striking distance of the combined Japanese fleet. Luring the strategically superior power into the teeth of a regional defense would appear a significant anti-access tactic. In his study of the Magic project, which decoded Japanese diplomatic cables, and of the U.S. signals intelligence and code-breaking efforts in the Pacific (also referred to as Ultra), Ronald Lewin maintains that the Imperial Japanese high command also "saw Midway and its surrounding seas as the remaining gap in the perimeter."[19] Capture of the Aleutians was merely a diversion, and although control of Midway could give the Japanese a base for further operations against Hawaii, it was most valuable in solidifying the imperial wall at sea.

Unfortunately for Japan and Yamamoto—who also planned the Midway operation—the United States was not only strategically superior, but also far superior in its cryptologic and signals intelligence capability. Not only had the United States been reading Japan's "Purple" (diplomatic) codes, but it had begun to break Japanese naval (and later military) codes by the spring of 1942. SIGINT intercepts were also well analyzed, although there was a minor degree of success for occasional Japanese deception. The Japanese navy was adept at operating in radio silence. But at the time of Yamamoto's movement toward Midway, the commander in chief of the U.S. Pacific Fleet, Adm. Chester W. Nimitz, had a reasonable knowledge of Yamamoto's objectives, the general composition of his force, and the approximate date of the operation. Good admiralship was required to win the battle, but intelligence superiority gave the United States considerable, perhaps decisive advantages.

Following the Battle of Midway, Allied operations in the Pacific War consisted of chipping away at the Japanese anti-access wall island by island, with amphibious assaults as the primary operational tool, which in turn enabled the capture of land bases for Army Air Force bombing of Japan. Also of great importance was the Allied submarine war against Japanese shipping, which denied the transit of resources from captured areas to Japan. American submarines can be considered the first counter–anti-access force to penetrate Japanese defenses.

Assessing the Pacific War: Fundamentals

From the very beginning of the Pacific War, thoughtful Japanese strategists such as Yamamoto recognized that the United States was a *strategically superior* (1) power that could potentially interfere in the continuing war with China and reverse expected gains in Southeast Asia and the Dutch East Indies. But blinded by their sense of racial superiority, other senior Japanese army officers did not concur with this assessment. Any conflict with U.S. forces would be largely a defensive naval engagement, a fact that elicited a sense of some disinterest on the part of an army leadership fixated on potential conquests in China, the Philippines, Indochina, Malaya, and the Dutch East Indies. The combination of racism, focus on the offensive, indifference toward maritime operations, and absolute commitment to expanding the empire led the Japanese army–controlled government to set a course on inevitable war with the United States.

The point of repeating these failings in Japanese assessments is to illustrate the inability of the United States to effectively deter Japanese actions. And the point of recalling the U.S. inability to deter the Pacific War (and to some extent, the inability of Poland, France, and Britain to deter the European War) is to point to the fact that overall strategic superiority in itself is not necessarily a deterrent to aggressive actions by a force that views itself largely immune from counterattack. An apparent ability to penetrate regional anti-access defenses would appear a more effective deterrent. Recalling the complaint by the Japanese army planner, faith—however mistaken—in the strength of anti-access defense can lead to the acceptance of considerable risk when facing a strategically superior opponent and to precipitous actions. The American deterrence effort—and President Franklin D. Roosevelt had ordered the U.S. Navy to shift its capital ships from the West Coast to Hawaii as a deterrent—failed because the dominant Japanese decision-makers did not believe that the United States was willing to make the sacrifices in blood and treasure needed to crack their anti-access network. Obviously they were wrong.

Whatever Japanese army leaders believed, Yamamoto and other naval leaders calculated that the United States would not be able to conduct a major maritime offensive until 1943 and that a victory in his Midway operations would be the decisive battle in that it would cause U.S. decision-makers to conclude that breaking the Western Pacific wall was too difficult. But he, too, was wrong.

It is obvious that the *geographic characteristics* (2) of the Asia-Pacific controlled the nature of the war, necessitating the perfection of such difficult operations as amphibious assault and combined carrier group strikes. Contrary to the Japanese army focus on the China theater, the maritime domain was the *predominant medium* (3) of conflict.

The *criticality of intelligence and information* (4) has already been noted. Both sides attempted considerable use of deception, but U.S. SIGINT and cryptologic efforts proved far superior. This was also enhanced by the extreme complexity of Japanese operational planning, often requiring the combination of disaggregated forces operating with little room for error . Keeping forces disaggregated was meant to be a tool of deception to prevent the Americans from ascertaining Japanese objectives, but the coordination required and the dependency on achieving surprise resulted in significant failures even when U.S. intelligence was lacking. Having constructed a wall around the western Pacific, the Japanese could not match the long-term industrial and technological superiority of the United States and its allies, and keep the anti-access network intact.

The U.S. Navy also had a number of successes in deception. Perhaps the most significant was the radio traffic simulation of a carrier task force in the southwestern Pacific on the eve of the Battle of Midway, when in fact the actual task force was within the fleet poised to meet the Japanese.[20]

Clearly the Japanese intended their alliance with Nazi Germany and Fascist Italy to generate the *extrinsic events* (5) that would prevent or distract the Allies from effectively challenging their "Greater East Asia Co-Prosperity Sphere." This did have the result of necessitating a Germany-first Allied strategy, but that choice would have likely been made in any event. In fact, it can be argued that the Tripartite Pact actually played a significant role in assuring an Allied victory in Europe—it allowed for the complete discrediting of the American isolationist movement and the U.S. decision to enter the war against Germany. Both the Japanese effort to present the United States with distracting extrinsic events by its alliance with Germany, and the German attempt to do likewise by declaring a simultaneous war against America, actually had disastrous results for their respective objectives. The key to American persistence against strong anti-access defenses was a total commitment to victory enhanced by the odious natures of the opposing regimes.

Assessing the Pacific War: Supplementary Factors

The Japanese achieved both *strikes against regional bases* (A) and *preemption* (B) with initial tactical success. Nearly simultaneous strikes on the Philippines and Hawaii along with follow-on thrusts into Malaya, the Dutch East Indies, and even the Indian Ocean removed all Allied regional bases in the opening phase of the war. As concerns the European colonial powers—Britain, France, and the Netherlands—obvious *extrinsic events* (5), the war in Europe, greatly facilitated these efforts. In the case of the United States, the inherent public pressure in a democracy for a peaceful resolution of differences and a flawed deterrence strategy enabled Japanese tactical success, along with the very great distances involved for the transit of forces from the continental United States. But *strategic superiority* (1) and an innovative culture proved decisive despite setbacks.

Japan had conducted significant *technological innovation* (C) prior to the war. In addition to long-range military research, Japanese scientists and technicians proved adept at solving tactical problems. One example is the effort to develop a shallow-running torpedo that could be dropped by aircraft and operate in the forty-foot depth of Pearl Harbor. This was finally perfected only three months before the attack.[21] One of the reasons the defenses of Hawaii were ill prepared was the U.S. Navy's belief that an attack would not be successful in the shallow waters of Pearl Harbor. Shallow-running torpedoes were a technological surprise.

Likewise, an innovative solution to penetrating the armored decks of the moored battleships in the aerial attack was to convert 16-inch armor-piercing naval shells into bombs—which today might be considered sidewise technology.

What hindered Japanese technical innovation was the lack of industrial production capacity to turn innovations into large quantities of weapons, along with the Japanese islands' lack of the required raw materials. Technological innovations, such as the Long Lance heavy torpedo, could facilitate tactical innovations such as night surface attacks. But, as Yamamoto maintained, Japan could not win a war of matériel attrition.

Perhaps the most glaring deficiency of the Japanese war machine was its routine inability to achieve effective *cross-domain synergy* (D). In many warfare areas, Japanese tactical capabilities matched the Allies. Carrier strike tactics paralleled those of the Americans, although they were weakened by the relatively low interest in mastering carrier damage control. In land war, the Japanese fought with a ferocity intended to make up for inferior numbers. But despite the fact that Japanese naval commanders utilized tactical land-based aviation to considerable extent in the southwestern Pacific area (in fact, they relied on it), cross-domain synergy continuously eluded them. Even in naval combat alone, undersea forces were poorly coordinated with surface and air forces, with submarines consistently out of position during attempts to support major engagements (most notably at Midway).

Overall army-navy cooperation was limited to individual initiatives. Mostly there was great hostility between officers of the two services. A significant factor was that not only were they rivals for defense resources, they were also rivals for political power with both army and navy flag officers fighting for the post of prime minister. Prior to actual hostilities, the army considered naval leaders to be pro-American defeatists, and some of the prewar assassinations or attempted assassinations of officials (including Yamamoto) were plotted by army officers. Noncombat units, such a cryptologists and intelligence officers, also appeared to be held in relatively low regard. Under the circumstances, the cross-domain synergy that might occasionally be achieved was fortuitous or miraculous.

In contrast, the U.S. joint chiefs did not allow service rivalries to develop—despite the fact that they viewed Gen. Douglas MacArthur as a discordant element to be tolerated owing to political considerations.[22]

Cross-domain synergy may have been limited in comparison to modern operations, but the linkage was consistently attempted. The fact that many operations were exclusive to the Navy and Marine Corps, both naval services, ensured a common doctrinal base. However, both the Army and Navy strove for coordination, and both regarded logistical and support services as essential, not secondary. Both relied on intelligence units and shared resources in that regard. Naval analysis of the coded messages of the Japanese ambassador in Berlin and in other European capitals helped confirm European Ultra intelligence, and Army cryptologists later took responsibility for Magic to allow the Navy to focus on breaking the Japanese naval code. Intelligence policies were controlled by a joint intelligence committee.

In assessing the Pacific War as an anti-access vs. counter–anti-access struggle, the achievement of information and intelligence superiority, consistent efforts at achieving cross-domain synergy, and an absolute commitment that made extrinsic events irrelevant were all critical in allowing the counter–anti-access force to bring its strategic superiority to bear relentlessly on the enemy.

Case 3: The Struggle to Retake the Falklands

The other cases we have examined occurred in the context of global war. The Falklands case identifies the causal factors that can propel a regional state to militarily challenge a strategically superior power under unique and isolated conditions.[23] Additionally, it illustrates how a modern conflict can occur between an authoritarian government and a democratic state. Argentina was then ruled by a military junta at war with many of its own citizens.

Argentine forces had invaded and annexed the territory of a nuclear-armed, traditionally global power with greater economic and diplomatic influence. Such a challenge logically involves an anti-access phase if a countereffort occurs. The United Kingdom succeeded in the countereffort.

Events of the Conflict

Whatever the tortuous legal history of the possession of the Falklands/Malvinas, it was clear that the islands' inhabitants considered themselves

British and firmly desired that the territory remain part of the United Kingdom as it had been for nearly 150 years. Early in the previous century, de facto control of the islands changed several times—even U.S. Marines were briefly in control (in 1832).[24] By 1982, claims by the Argentine government that the British presence represented colonialism no longer seemed logical. Rumors of undersea oil seemed to make the islands a more valuable commodity, but many observers assumed that tapping such wealth might be a joint endeavor. The obvious question is why the Argentine government decided to use preemptive force in seizing the islands.

The answer lies in Argentina's domestic situation at that time. The government was in the hands of a military junta that had ousted the populist president Juan Perón in 1976, ostensibly to avoid economic collapse. But by the sixth year of military rule, the economy was characterized by "falling industrial output, falling wages, rising unemployment, [and] a massive inflation rate above 100 percent and still climbing."[25] In addition, the junta conducted an internal "dirty war" against political opponents that included torture, disappearances, and at least seven thousand citizens killed (perhaps as many as fifteen thousand).[26] With no appearance of legitimacy, the junta resorted to resolving a standing nationalist grievance, the outcome of which could appeal to Argentine patriotism and popularly justify continuation of junta rule.

The invasion of 2 April 1982 was spearheaded by the elite Buzo Tácticos marine commando unit, with a total of seven thousand Argentine troops facing a defense force of sixty-eight Royal Marines. With those odds, capture of the islands went rather swiftly following an administrative landing—there was no need for an amphibious assault (although they were shot at by a two-marine machine gun team). The Argentine junta had assumed that the fait accompli would lead to the United Kingdom's acknowledgement of Argentine sovereignty, perhaps after a face-saving diplomatic maneuver. The launching of a hundred-ship British task force to make the 8,000-mile (12,875-kilometer) voyage to retake the islands was clearly unexpected.[27]

By 4 April the British nuclear-powered attack submarine HMS *Conqueror* left Scotland as the lead for the U.K. task force. (The *Daily Telegraph* reported that another submarine had previously sortied on 30 March.)[28] On 5 April the first units of the surface force, which was included the aircraft carriers HMS *Hermes* and HMS *Invincible*, started the voyage from their bases in Britain. Both could accommodate only helicopters and V/STOL (vertical and/or short takeoff and landing) fixed-wing aircraft. Naval ships and auxiliary vessels carried approximately 5,000 marines, paras, and infantry troops. The total number of British airman, marines, sailors, and soldiers involved in the operation was about 28,000, although most of that number remained at sea or performed support functions. British forces landed ashore were actually inferior in numbers to the 12,000 Argentine troops defending the islands—although not in training and commitment. Many of the Argentines were conscripts, and their commanders had not anticipated having to mount a determined defense.

On 12 April the United Kingdom declared an exclusion zone of two hundred nautical miles around the Falklands, within which Argentine forces would be attacked. A small British force recaptured South Georgia, a British dependency east of the Falklands and uninhabited except for a scientific team, on 25 April. The British assault there also damaged an Argentine submarine. On 2 May HMS *Conqueror* sank the Argentine cruiser *General Belgrano*, which had been steaming to the southwest of the islands but outside the exclusion zone. This attack energized so-called antiwar groups in Britain and elsewhere, which claimed that *Belgrano* had been leaving the Falklands area. However, reconstructions indicate the opposite, and it was equipped with the surface-version of the Exocet antiship missiles that were to prove the most damaging weapon against the Royal Navy when launched by aircraft.[29] The immediate effect was that all other Argentine surface ships returned to port and remained there for the rest of the conflict. Ironically, several of the Argentine warships were of the same class as the Royal Navy destroyers. (The Argentine submarine *San Luis* did fire six torpedoes against apparent British targets but without success.[30])

By 5 May many of the British ships were arriving in the area. Argentine forces unleashed their most potent weapons, tactical attack aircraft flying from bases in Argentina armed with Exocet missiles as well as bombs. The first British warship to be sunk was the destroyer HMS *Sheffield*. A total of six British warships and one commercial vessel converted as an aircraft transport were sunk in the conflict. However, the Argentine air force actually failed in its primary mission: locating and attacking the two British aircraft carriers. Argentine aircraft also failed to attack the amphibious shipping. Part of the reason is that the aircraft generally attacked the first target encountered, which included the picket line of destroyers and frigates meant to defend against air attack. This was a trait similar to the kamikaze attacks of the Pacific War, when destroyers took the brunt of the attacks while high-value units, such as carriers, were not effectively threatened.[31]

British landings at San Carlos on the side of East Falklands opposite the sole island village of Stanley, the site of primary Argentine defenses, commenced on 21 May. Argentine air attacks failed to interfere. Argentine forces had not planted sea mines in the area, which was fortunate for the British because their task force had no countermine vessels or aircraft. In order to ensure there were no mines in the amphibious operations area, the frigate HMS *Alacrity* was sent to simply sail through the littoral and see if any mines were set off.

Once British troops were ashore, their superior training and conditioning led to a series of victorious engagements as they marched across the island. The eventual encirclement of Stanley resulted in the surrender of Argentine forces on the Falklands by 14 June. In total, 255 British service members and approximately 800 Argentine were killed. Together somewhere around 2,000 combatants were wounded. British units documented a total of 12,978 Argentine prisoners of war, all of whom were repatriated to the mainland.[32] The defeat in the Falklands caused the fall of the military junta.

Anti-access Systems and Tactics

Although the Falklands War seems a "minor" conflict in comparison to the other examples, modern anti-access systems and tactics were utilized by the Argentines, albeit with poor effect. Argentine forces did not have the capability to conduct a regional defense and keep the British outside of the South Atlantic. They could not strike at U.K. regional "bases" (such as they were), primarily Ascension Island, which was 3,300 miles (5,311 kilometers) from the Falklands. But their most effective weapons, air- and surface-launched antiship cruise missiles, would be a tool of any anti-access defense at sea. Land-based attack aviation was the primary Argentine anti-access combatant force, similar as it would be in larger engagements. Significantly overmatched, the Argentine command decided to withhold the surface naval forces that might normally be significant in sea denial—probably a very wise decision. Submarine attacks on the task force were attempted, but ship for ship the Royal Navy had the world's most skilled antisubmarine watch teams. Argentina had no nuclear-powered submarines, and the operating distances stretched the capacity of diesel submarines.

A significant British advantage was long-range and satellite surveillance, bolstered by American assistance—capabilities Argentina did not possess even as a regional power close to the combat zone.

Argentina failed to use some anti-access systems and tactics within its capabilities. It made no use of sea mines, which would have been the most substantial threat to an amphibious landing. The probable reason was that the British response was unexpected and Argentine forces had limited time to sow naval minefields. Prewar Argentine diplomacy was unable to influence world opinion, both because its case was weak and because states of interest considered the junta odious. Argentina had similar territorial disputes with neighboring Chile, which may have also provided intelligence to the British. The Organization of American States (OAS) voted in Argentina's favor, but only Peru gave active assistance—although some weapons may have come from the Soviet Union via Libya and Brazil.[33] If so, they were too late or not of significance.

But the ineffectiveness of their efforts do not invalidate the reality that—in the face of the qualitative advantages held by British forces—the only chance of Argentine success was to keep the task force away from the islands until time, attrition, and/or extrinsic events shook the determination of the attacker.

Assessing the Falklands War: Fundamentals

Despite the waning of her global power and years of economic difficulties, Britain was *strategically superior* (1) to Argentina. The very ability to mount a joint force expedition over 8,000 miles (12,875 kilometers) is evidence of that fact. Despite contemporary economic woes, the British economy and industrial base were superior to those of Argentina in every measure. Its standing in the world may not have been what it was in the days of its colonial empire, but its global influence far outstripped Argentina's. The United Kingdom was/is a nuclear power allied by treaty to NATO nations, including the United States. U.S. commitment to the

OAS, of which Argentina is a member, has never come close to its commitment to NATO.

Geographic characteristics (2) were obvious factors, the most significant being the great distance from the United Kingdom and from British bases. Another often overlooked factor was the coming of winter with its fierce winds whipping across the treeless Falklands. It is possible that if the Argentine defense had managed to last but a few more weeks, the weather could have forestalled the British advance on land, as well as at-sea combat operations. This was a climatic anti-access element of which Argentine forces did not take advantage.

Although a ground campaign was necessary to recapture the islands, the *maritime domain* obviously remained the *predominant medium* (3) on which all operations were dependent. As First Sea Lord Admiral Sir Henry Leach described the circumstances, "there was no conceivable means of any U.K. agency doing anything about it [the Argentine invasion] unless they were got there by the Navy and protected by the Navy when they got there."[34] This dependency was reflected in the fact that the United Kingdom had to rely exclusively on carrier or other ship-based aircraft, giving Argentine forces a potential advantage. But the cross-maritime distance forced the Argentine land-based aircraft to operate at the extreme range of their fuel endurance.

Intelligence and information (4) remained critical, although the advantages were all on the side of the United Kingdom. It is evident that the Argentine command on the mainland did not have solid situational awareness about what was happening on the islands. On the other side, information, situational awareness, and communications were such that operational direction could be given from the United Kingdom. HMS *Conqueror* waited for a decision by the British Cabinet before it torpedoed *Belgrano*, immediately starting its attack sequence after the decision was made in London.[35]

Argentina continually hoped that *extrinsic events* (5)—primarily diplomatic—would interfere or delay Britain's response. In reality, the primary element most likely to affect British decision was the financial cost of the operation and its effect on the U.K. economy. Prime Minister Margaret Thatcher's government had been elected to solve economic problems, and predecessor governments had largely written off the faraway Falkland Islands and its sheep as an economic liability not worth the cost of a truly effective defense.[36] But Thatcher—indeed most all the British—became adamant that the military aggression would be reversed. The Argentine president, army general Leopoldo Galtieri, firmly believed that the relationship he had cultivated with the United States in previous years would cause it to either support Argentina or simply elect not to furnish aid to the British. But, again, he was mistaken. Once the United Kingdom decided on its course, no extrinsic events occurred to sway it from its heading.

Assessing the Falklands War: Supplementary Factors

Argentine forces did not *strike regional bases* (A) of the United Kingdom because there simply were no regional bases to strike. Along with the Falkland Islands, Argentine forces had taken South Georgia, but it was not a location that could support military forces. The nearest British territories of Saint Helena and Tristan da Cunha were not military bases. Ascension Island was a way station that could support communications and intelligence traffic and limited long-range aircraft, but it too could not shelter a fleet and was far out of range. U.K. forces were completely dependent on forming a sea base of ships. Of course, striking this sea base was the main operational objective of Argentine forces.

Obviously Argentine forces used *preemption* (B), an attack that was the very cause of the war. But the preemptive attack was on the territorial objective, not on the British task force then 8,000 miles (12,875 kilometers) away.

In the form of Exocet missiles—which had not previously been proven in combat—the conflict did include *technical innovations* (C). But these innovations were not specifically tailored for anti-access operations per se. Technical innovations were used on both sides. The overall quality of British forces simply proved superior in all regards.

Finally, the British forces can be considered to have achieved a degree of *cross-domain synergy* (D), whereas the Argentine forces could not. In fact, Argentina could not reinforce or resupply its ground forces on the Falklands once actual combat operations commenced. Aircraft attacks were not coordinated with ground operations, particularly in opposing a British landing. Argentine ground forces scored a hit with a land-based Exocet on HMS *Glamorgan*, but that too did not represent a coordinated attack. Britain had the capacity to achieve cross-domain synergy; Argentina did not.

Lessons of the Counter–Anti-Access Campaigns

Common strands appear to run throughout the three examples of anti-access defeats that can provide lessons for future counter–anti-access campaigns. First, of course, is the realization that the fundamentals that define the anti-access situation were present. They are most likely to be present in engagements with which the United States might be faced in the future. Secondly, preemption is a definite historical factor in both achieving aggressive gains *and* buttressing an overall anti-access campaign. Anti-access campaigns are dependent on clearing out enemy forces already in the region in order to focus on stopping forces outside the region from reentering. If opposing forces are *not* cleared from the region, the out-of-area power retains a toehold on which it can leverage its power-projection force. If they are cleared, the power-projection forces are required to face the time and attrition necessary to construct a sea base, build or obtain new facilities in allied states, or strike at longer ranges.

Information, intelligence, and deception are all keys to potential success by both the anti-access and counter–anti-access forces. But in some regard they can prove to be of greater advantage to counter–anti-access forces because they can approach the battle space on multiple axes, whereas the general location of anti-access systems are likely to be initially known. Deception is critical. Hitler did not know the location at which the Allied landing would occur, was forced to stretch his anti-access resources, and was deceived as to the Allies' primary thrust. The Japanese knew the general geographic points that the Allied forces would need to capture or traverse to reach the home islands. But they were tactically deceived in key battles and were deceived in believing that their communication codes were secure. Argentine air forces were never able to locate the British V/STOL carriers, but the British certainly knew where the bulk of Argentine land forces needed to be. They could initially work around them and then surround them.

Technological superiority played a lesser role in the cases because the degree of superiority was never so great as to be decisive. Production capability of weapons and forces *did* prove decisive, but the technologies of war appear to have been diffused. This may be even more real today as information on technology travels at lightning speed and can be duplicated. Industrial base may be more important than technological base—a thought that is not currently in vogue.

The forces that achieved a greater degree of cross-domain synergy were indeed victorious, but it must be admitted that such is the case in all combined arms warfare. Militaries that can obtain cross-domain synergy are simply better, more capable militaries. This observation is not meant to denigrate jointness or the overall need for cross-domain synergy. It simply points to the reality that cross-domain synergy should be achieved whether facing an anti-access network or not. It should always be a prerequisite to military operations.

In all cases, it was an absolute commitment to victory that enabled the counter–anti-access force to utilize its strategic superiority. Extrinsic events were ignored or simply were not significant enough to merit a pause in operations or reconsideration of ultimate objective. All three anti-access defenses were predicated on some degree of support from the distracting effects of extrinsic events. Hitler hoped that Japanese successes in the Pacific War would divert American resources away from intervention in the European theater. Imperial Japanese leaders hoped that involvement in Europe would bleed American strength. Argentina assumed that diplomacy and the financial costs of recovering "islands most Britons could not locate on a map" would cause the U.K. government to bow to the fait accompli. In addition, Galtieri thought that he had built a relationship with Washington—based on common Cold War concerns—that would cause it to dissuade the Thatcher government from action or at least not support it.

It must also be noted that in all cases, initial deterrence efforts by the strategically superior power failed. In the cases of the Pacific War and Falklands War, the strategically superior power's military forces in the region (such as American forces in the Philippines) were simply judged too small to interfere with the enemy's objectives. They could barely defend themselves, let alone break through to conduct offensive operations. President Franklin Roosevelt's decision to move the capital ships of the U.S. fleet to Pearl Harbor did not deter Japanese plans—as he had hoped—because Pearl Harbor was viewed as too far away from the western Pacific to allow U.S. forces to quickly interpose themselves in the areas of Japanese conquest and project power into the region. Moreover, concentrating the fleet at Pearl Harbor, a location more difficult to defend from attack from the many quadrants of the open sea than the California harbors of San Pedro/Long Beach or San Diego, made the fleet a tempting target for preemption by a determined opponent. It did the opposite of deterrence: it allowed strategic skeptics of the Japanese decision for war to convince themselves that—with luck—a prolonged war leading to eventual defeat could be avoided. As Yamamoto noted about himself when discussing Pearl Harbor plans with internal navy critics, "I like games of chance. You have told me that the operation was a gamble, so I shall carry it out."[37]

Power-projection forces of the strategically superior power were also deemed to be impeded by distance. In the case of the Japanese, attrition attacks against U.S. forces as they steamed their way across the Pacific had been part of their war plan for over twenty years. They were convinced that their anti-access network would hold firm once the U.S. fleet was struck by the opening preemptive blow. Remnants of that fleet would be destroyed in a decisive battle closer to Japanese waters. The Argentine junta did not plan—or have the capability—to strike at the Royal Navy as it entered the South Atlantic region. But it did expect the distance involved to be a deterrent—a prime element of an anti-access posture that was dependent on geographic characteristics. The Argentines also expected greater effectiveness from their air force in striking the British carriers, their navy in engaging surface units, and their army on the islands in preventing an amphibious landing. All three expectations were miscalculations, but with luck such actions could have delayed the British until—per one of the repetitive themes of this book—time, attrition, and/or extrinsic events shook the determination of the attacker.

In the case of Nazi Germany, its leaders assumed that their genius in tactical innovation (the blitzkrieg) along with technological advances (aircraft, tanks, and eventually long-range missiles) would make up for any Allied superiority in numbers. Unlike the Japanese, the Germans did not have utter disdain for the fighting spirit of their opponents, although Hitler viewed western Allied leaders as weak and indecisive, and Stalin as internally vulnerable. But like the Japanese, he did not expect his opponents to have the same will to fight that he possessed.

The Allies' potential strategic superiority once the United States entered the war, along with the resupply capabilities of the dominions of the British Empire (though hazarded by unrestricted submarine warfare), simply did not deter him. Even if the Allies commenced offensive (power-projection) operations, Fortress Europe was expected to prevent their return to the continent.

From a modern perspective, these failures in deterrence are disturbing. Also disturbing is the dominant use of preemption to try to prevent the opponent from using its strategic strength to reenter the region or to dissuade it from power-projection operations that are "just not worth it" given the possible costs. In this the attacker or eventual anti-access defender misinterpreted the conflicts to be "wars of choice" for the strategically superior, out-of-region power, rather than "wars of commitment." One must ask whether this miscalculation is common to authoritarian states, particularly if they view their objectives as "limited."

Would these states have gone to war if they had been convinced that their out-of-area opponent had the capability and will to defeat their anti-access defenses, to break their great walls? That of course remains unknown. But it is logical to think that if their assessment of the "correlation of forces"—an apt term used by the Soviet military—had included the conclusion that their opponent could eventually break through their anti-access network and threaten not just their gains but their overall military strength (and perhaps internal power control), they would have been deterred to a greater degree.

This view points to the need for adding specific and robust counter–anti-access capabilities and planning to one's military structure. It also brings us back to the discussion of the deterrent effect of counter–anti-access capabilities. Judging by these historical cases, it will not be enough to be theoretically able to defeat an aggressor in the "fair fight" of a force-on-force engagement. Achieving a condition of conventional deterrence in an anti-access environment requires more. In an age of expanding technological capabilities that can enhance anti-access networks, what appears to be required is the specific capability to break down, deceive, circumvent, or otherwise neutralize anti-access defenses in order to convince the defender to forgo thoughts of expansion or "redress of grievances" by force. This requires continuing the nascent efforts to include the counter–anti-access campaign as an intricate and specific phase of joint military operations in event of war. This may also require the tailoring of force structure so that the forces most needed in creating breaks in the great walls get a greater proportional share of defense resources. Such a force restructuring may not yet require additional resources added to the defense budget. Rather, it may require a judicious reallocation of resources. This may not correspond with the preferred view of those who see land forces as the primary arm of joint operations. Indeed, land forces are necessary if one wishes to invest, decisively defeat, cause a definite regime change, or pacify an opposing state. But land forces that cannot enter the region of conflict—particularly in regions defined by maritime boundaries, fewer or vulnerable regional land bases, or geographic conditions that facilitate an anti-access defense—do not seem of primary value.

Also needed, of course, is the will to commit to the fight despite opposing pressures.

Chapter 6

East Asia: Most Formidable Challenge

Moving away from the historical examples, we enter the realm of estimates and speculation. But all planning is to some degree speculation. Events do not always coincide with our plans. Prussian field marshal Helmuth von Moltke the Elder famously wrote, "No plan survives contact with the enemy."[1] More recently, former secretary of the Navy Richard Danzig argued that "to the extent we foresee the future and effectively address it, then the future will not develop as we anticipated it."[2] Yet planning is essential for peace as well as war. Deterrence is planning for war in order to keep the peace. The first role of counter–anti-access planning is deterrence.

It is with this in mind that we begin the next few chapters, which individually describe a particular anti-access scenario that could face U.S. armed forces, allies, and partners in the near future. This is not an attempt to predict what *will* happen, even in an anti-access campaign, but to identify what *might* happen given military and political trends.[3] More importantly from the U.S. perspective is the identification of actions that can be taken to counter anti-access actions that may occur. This will be done in the context of the fundamental elements and supplemental factors used in analyzing the historical examples.

These chapters will not attempt to duplicate the extensive work already done, especially by CSBA, on the details of the two most significant scenarios: probable anti-access strategies of the People's Republic of China and the Islamic Republic of Iran. Many other military officers and scholars have contributed much to the public understanding of what might take place. Rather, the attempt is to supplement them by the use of the methods of this book.

Additionally, current orders of battle or details of specific weapon and sensor systems are not included. They are continually changing as new systems are developed or adopted during the constant evolution in military affairs. Up-to-date information on these is readily obtainable elsewhere. The concentration here is on the fundamentals of anti-access.

Before beginning, another comment needs to be made about the purpose of these scenarios. Even more recently—as described in Chapter 2—a retired general expressed his discomfort with the existence of the AirSea Battle concept and the study of the anti-access attributes of the PLA by saying that "Air/Sea Battle is demonizing China. That's not in anybody's interest."[4] The reality is that no examination of a potential war strategy can demonize the Chinese Communist Party any more than the CCP has demonized itself throughout its, shall we say, conflictive history. The CCP rather than China is referred to here because remarkably unacknowledged in most discussions of China's political future is the fact that the People's Liberation Army is pledged to the party, not to the state. Technically, China the nation has no armed forces. With a Western-centric view, some would dismiss this as mere nuance. However, it has a direct influence over the possible causes of conflict with the United States and its allies and partners. Demonizing or not, it is prudent to examine all aspects of potential events that *might occur*. And if they might occur, it is very prudent to examine *how to deter them*. If they can't be deterred, then it is critical to discuss what must or should be done.[5] If Danzig is right, then addressing these issues might lead to a very positive outcome.

Another reason that it is valid to examine anti-access strategies vis-à-vis counter-anti-access strategies in an East Asia scenario is that there is ample evidence that PLA strategic thinkers envision the very same scenario as the most likely "local war under high-technology conditions" that they would face. The most likely opponent, of course, is the United States. As noted, PLA commentators do not use the term *anti-access* (and it may not even be directly translatable), but the operations they describe and sometimes advocate are largely the same as have been described throughout this book as supporting anti-access strategies.

Anti-Access Systems

More important, the PLA appears to be acquiring weapon systems that are optimized for an anti-access strategy and particularly for attacking regional bases where U.S. forces might be stationed. Significant resources have been poured into developing ballistic missiles of various ranges, and the status of the Second Artillery Corps has risen in the PLA since its creation in the late 1960s as the nuclear weapons force. Today it controls all land-based conventional missiles (ballistic and cruise missiles) as well. In 1995 and 1996, Second Artillery missiles were test-fired into the sea within a hundred miles of Taiwan. As noted in Chapter 3, the PRC has also concentrated on developing and improving its inventory of intermediate-range ballistic missiles, a category not in the U.S. inventory because they were eliminated in 1980s arms control agreements with the Soviet Union. Ballistic missiles are not inherently defensive weapons.

ASAT testing demonstrates a capability for blinding an opponent's satellite reconnaissance and surveillance assets and possibly communications. In Chapter 3 this was identified as a necessary initial step in anti-access operations. Additionally, many U.S. precision weapons are guided by Global Positioning System (GPS) data, and destruction of GPS satellites may affect the successful targeting of such weapons.

At the same time as it is developing antisatellite weapons, the PRC is developing a new range of detection and surveillance (and possibly targeting) satellites. A number of recent satellites have been described as "navigational" or "environmental sensors." It is highly possible that these satellites, particularly those that have ocean observation or radar capabilities, actually have military missions.[6] PRC transoceanic shipping can easily rely on existing commercial satellites for navigational purposes. Environmental information is also available from commercial sources. Gathering such information via dedicated satellites can certainly have dual use. The PRC appears not to have yet duplicated the Soviet Cold War naval surveillance satellites such as the Electronic Ocean Reconnaissance Satellites (EORSATs) or Radar Ocean Reconnaissance Satellites (RORSATs) that could range globally and be steered in flight to detect carrier battle groups, but such would be a desirable capability. Supporting the satellite-detection capabilities are networks of land-based OTH-B radars designed to detect vessels beyond the normal radar horizon.

The purpose of ocean-surveillance satellites is to complement the construction program to create a sea-denial navy that can reach at least the first of two notional defensive sea perimeters, the island chains that have been discussed by some PRC commentators. The first chain runs from the southern tip of the Kamchatka Peninsula through the Japanese Islands, Taiwan, the Philippines, Malaysia, Brunei, and Singapore. The second island chain runs from Japan through Guam, the Marianas, and Micronesia to northern Papua New Guinea. It is interesting that these island chains are officially described in PRC military literature as barriers to Chinese access to the open ocean, yet are described in the same literature as being a defensive concept.[7] Since the PRC has yet to develop a fully blue water navy that has the logistical capability to support itself for long periods outside the home region, it seems that its study of the island chains is intended to benefit further development of a longer-range anti-access network rather than plans to "break out" from naval encirclement. Some of its potential regional military opponents lie within these island chains. Bernard Cole of the U.S. National Defense University has described the developing People's Liberation Army Navy (PLAN) as constituting "the great wall at sea."[8] The concept of the island chains as barriers that could be defended to prevent U.S. naval and joint power-projection forces from entering into the region should be considered a part of this great wall. Details of PLAN naval developments have been extensively analyzed in works by scholars associated with the China Maritime Studies Institute of the U.S. Naval War College.[9]

China has continued to add density to its air defense networks and extend their reach. A program of continuous improvement is apparent. In May 2012, Voice of Russia radio announced that the PRC intended to buy an unspecified number of Russian mobile S-400 Triumf air defense systems (NATO reporting name SA-21 Growler).[10] The S-400 system combines missiles of three different ranges to achieve a coverage range to 250 nautical miles. Many analysts consider it superior to the U.S. Patriot PAC3 system and perhaps the best air defense system in the world, and until recently the Russian ministry of defense had stated it would not be exported. It was not until 2007 that Russia itself deployed the system, which began development in the 1990s. Its reputation is such that South Korea, a U.S. military ally, is constructing its own smaller version in collaboration with the Russian defense firm Almaz, Samsung, and the European defense company Thales.[11] Saudi Arabia—another major U.S. arms client—is also reported to be interested in the system.[12]

There were reports that China had developed an S-400 variant jointly with Russia in 2009, designated HQ-19.[13] However, this may be primarily a version of the S-300, an earlier air defense system employed by China. In Russia the S-400 is replacing the S-300. Although initially deployed by the Soviet Union in 1979, the S-300 is itself a formidable air defense system that has been upgraded, its current version being the S-300PMU2 (designated HQ-18 in the PRC).

Returning to the issue of conventional ballistic missiles, development of antiship ballistic missiles by the PRC has been previously mentioned. This has become a controversial subject in open, unofficial U.S. defense literature. Reputable authorities maintain that ASBMs have become a major component of the PRC anti-access network, to primarily target U.S. aircraft carriers operation in the western Pacific.[14] The ability to hit such moving targets at sea is attributed to its payload of maneuvering reentry vehicles (MaRVs) with infrared (IR) seekers that could provide for midcourse targeting changes. Impetus for development of this particular weapon has been "traced in part to the 1995–96 Taiwan Strait crisis, which further underscored Chinese feelings of helplessness against American naval power," as Andrew S. Erickson and David D. Yang have written.[15] Some media reports have suggested that this development makes aircraft carriers obsolete, but others suggest that sound tactics can neutralize the ASBM threat.[16]

As is apparent, the lack of opaqueness in PRC military developments—a feature of CCP policy of which the U.S. government has frequently complained—makes it difficult to assess the actual capabilities of a Chinese anti-access network (at least not based on open sources). However, the development of anti-access capabilities appears unquestionable. Commentators concerned with possible "destabilizing" U.S. responses to these developments have suggested that such are merely "defensive measures" of a PRC naturally suspicious of the intentions of other nations. The problem is that the PRC, like Nazi Germany, Imperial

Japan, and Argentina under the junta, is an authoritarian state, and there is a major flash point for a PRC-induced conflict with the United States and its allies.

Cause of the Crisis

An East Asia scenario in which the People's Republic of China activates an anti-access campaign in support of a forcible annexation of Taiwan is the most formidable anti-access challenge that can be currently conceived. In their book of the same name, think-tankers Richard C. Bush and Michael E. O'Hanlon refer to a potential war with China as "a war like no other."[17] They also acknowledge that the most probable cause of a war between the PRC and the United States would be events concerning Taiwan; there simply do not seem to be any other issues that could cause a war between them. The status of South China Sea resources or the claims and counterclaims to ownership of other islands might create regional tensions, and incidents between naval ships and aircraft at sea could also cause considerable diplomatic and public furor, but the likelihood of such causing a direct armed conflict between the United States and the PRC is remote. Other factors, such as trade relationships, tend to mitigate small incidents. However, a military conquest of an island with a population of over 23 million or an attack on a U.S. vessel or aircraft in the vicinity of Taiwan would be another thing entirely.

Many argue for the improbability of such a war, and most of us hope they are right. The effect on China's export-driven economy is certainly an extrinsic event that should cause considerable deliberation in Beijing. Thomas P. M. Barnett, author of the best-selling *The Pentagon's New Map*, maintains, "fallout from a United States-China conflict over Taiwan would be enormous for globalization, effectively barring Beijing from stable Core [global economy] membership for the foreseeable future."[18] Yet there are troubling situations that can be contemplated. Contrarian strategist Ralph Peters advises, "Do not worry about a successful China. Worry about a failing China."[19] The logic of this viewpoint is that, if unable to deliver on its tacit agreement to deliver prosperity to the Chinese people in return for absolute power, the de facto legitimacy of CCP rule crumbles. This sets up a potential Falkland/Malvinas situation in which the ruling elite of the CCP can only retain the "mandate from heaven" by resolving a nationalistic, irredentist claim and "finishing the civil war" by taking Taiwan by force.

Since it would be a war for a limited objective, which by PLA calculations also might be limited in time, the anti-access objective would simply be to keep U.S. forces out of the region for that time. Aircraft carriers do not have to be sunk; it is enough that the United States would be reluctant to bring the carriers within range of PLA missiles. If a conquest or forced surrender of Taiwan—with its military capability destroyed through strikes by hundreds of conventionally armed ballistic missiles and land-based attack aviation—occurs rapidly, the PRC could settle into an anti-access posture in which American power-projection forces would expend their strength attempting to crack the great wall of a robust anti-access network. Even if U.S. forces could fissure the network, what force-on-force engagement could possibly reverse the gain? Would the United States be willing to conduct amphibious assaults on a PLA-occupied Taiwan? Would it be willing to engage in a land war in eastern Asia? Those are not simply scholarly speculations. They are calculations that CCP/PLA leaders may make if their power base is severely threatened and a dramatic act to redress "past wrongs" to China is needed.

The previous scenario postulates a "bolt-from-the-blue" attack in which the world has little warning as to final PRC decision-making concerning the use of force. The operational advantage of such surprise is that it could catch the majority of U.S. and allied forces in the region at a lowered state of readiness, with the bulk of CONUS-based forces in no position to respond rapidly. Even U.S. expeditionary forces would be hard pressed to land with Taiwan under attack, particularly when stationed in bases relocated from Okinawa to Guam—a thousand nautical miles farther away from Taiwan than Okinawa. S-400 air defense networks can reach out and beyond Taiwan's airspace. Combining land-based ASBMs and cruise missiles, much of the anti-access network would remain on the Chinese mainland while submarines, surface warships, and long-range aviation units try to block entry into waters beyond the island chains. This is not a complete parallel to the Japanese strategy against U.S. War Plan Orange, but the similarities are striking.

Preemptively striking regional bases where U.S. forces are located, such as fleet concentrations at Yokosuka, Japan, would remain a risky move. First, of course, a preemptive strike on U.S. forces could engage the Pearl Harbor syndrome, with the United States willing to go to extreme lengths to return to the western Pacific or, at the very least, use its economic strength to wreck a vulnerable Chinese export network. Second, the reaction of China's neighbors—particularly Japan—would likely be a permanent containing alliance of East Asian states, the encirclement that PRC strategists claim to fear.

An alternative scenario would be a gradual buildup in tensions resulting in a PRC-imposed economic embargo and physical blockade of Taiwan, combined with full activation of the PLA anti-access network. Although international law considers a blockade to be an act of war, the situation would force the United States into the position of starting active hostilities if it tried to force the blockade. The most significant extrinsic event would be global pressure to convince the United States to avoid the conflict at all costs lest it overturn the political and economic stability of the status quo. In any event, the United States would have to defeat a PRC anti-access network that would stretch over Taiwan in order to reverse PRC gains.

Anti-Access Actions

It requires no great degree of speculation to determine the actions the PRC would take in activating and operating its anti-access network. They have been outlined in general in Chapter 3, and CSBA, RAND, and others have described them in detail. What specifically sets the East Asia scenario apart from anti-access scenarios in other regions is the degree to which PLA strategists have considered the prospects and the amount of resources the PRC has been able to pour directly into anti-access capabilities. With the exception of a war with Russia, which the PRC apparently considers much less likely today than in the days of the Soviet Union, an anti-access–shaped conflict with the United States seems the only substantial challenge to PRC military capabilities and its huge PLA force. The fact that a force-on-force engagement on land—in which U.S. capabilities and training might match the PLA's advantage in the number of troops—is the least likely aspect of any conflict with the United States makes prioritization of resources into anti-access systems even more attractive. Likewise, the likelihood of an open-ocean clash (beyond the island chains) between the U.S. Navy and the PLAN is even more remote. This allows PRC resources to be channeled into a sea-denial navy with much the same mission as that of the Soviet navy: to keep the U.S. fleet away from near waters and out of position to conduct effective strikes ashore.

American military responses to the Taiwan scenario would depend on the initiation path for the conflict. The two broad categories are the *bolt-from-the-blue* (a) and the *gradual-escalation* (b) paths. Within each path are shaping characteristics that would determine the beginning phase of any U.S. response.

A bolt-from-the-blue/preemptive attack that is delivered on Taiwan but leaves U.S. and allied regional bases untouched—which will be referred to as *limited preemptive* (a1)—would allow for a much different response than would a full assault on U.S. forces located in the East Asia region, which can be referred to as *regionally preemptive* (a2). From an anti-access perspective, the regionally preemptive initiation provides considerable operational advantages and greatly enhances the strength of the anti-access network. Presumably the limited preemptive approach (a1) would be chosen in order to prevent collateral damage on the territories of U.S. allies and allow global diplomatic pressure to limit the scope of the U.S. response as well. Regional nations might want to avoid escalating a relatively local territorial conflict into a full regional conflict and might be reluctant to allow U.S. forces full access and use of regional bases. A hybrid approach might be an initiation phase that attempts to confine regional conflict to the maritime environment in order to bring the operational benefit of eliminating all U.S. naval forces operating in the region, yet avoiding damaged on allied territory. This would strengthen the resilience of the anti-access strategy, although not to the extent of a full regional preemptive attack (a2) on land and sea bases.

An unlikely alternative is a *long-range attack* (a3) on U.S. ports of embarkation for forces and logistics on the U.S. West Coast and Hawaii using conventional ballistic missiles. The strike would be focused on the sites of preparation for U.S. power-projection forces. While Chinese ballistic missiles can reach CONUS, their other force components cannot.

Gradual escalation (b) would allow nations opposed to the forcible annexation of Taiwan time to prepare to intervene. However, it would also allow for more complete use of political, diplomatic, and economic pressures by the PRC to generate or encourage extrinsic events that might blunt a response by the United States or others. It would also give opponents more time to consider if challenging an activated and alert PRC anti-access network is worth the costs. For democratic states, this would allow for the mobilization of so-called antiwar movements to question the need for a military response while negotiations might "prove effective." It would also allow those who prefer inaction to find justifying statements by Taiwan officials who choose to acquiesce to the mainland in the hope of retaining future political careers—or simply to avoid bloodshed. In a contest between democratic and authoritarian states, free public opinion is both a strength and weakness in the use of democratic power. If a gradual escalation scenario avoids recreating the Pearl Harbor syndrome, it might be a preferred path for the anti-access power with an offensive objective. Gradual escalation can be categorized by the use of *offensive anti-access tools* (b1), such as ballistic missile strikes or submarine attacks on warships entering the Taiwan Strait, or *relatively passive anti-access tools* (b2), such as emplacement of naval minefields around Taiwan in support of an embargo or campaign. If the crisis builds into a conflict, the anti-access state might still hope to limit the opponent's loss of life—primarily civilian—by confining attacks to the counter–anti-access power's long-range sensors, such as satellites. The tit-for-tat space war that might ensue could result in the destruction of satellites on both sides. This might prove an advantage to the counter–anti-access state, which could maneuver units across the Pacific unseen. However, it could distract attention away from events in the Strait, at least temporarily.

EMP generation is another technique that could be utilized to destroy sensors, including terrestrial and naval sensors. EMP generated by a nuclear burst can destroy many types of electronic circuitry, including that of computers, cell phones, and automotive starting systems. The obvious drawback is the introduction of nuclear weapons into the conflict, even if a high-altitude electromagnetic pulse (HEMP) detonation conducted above the earth's atmosphere would cause no direct personnel casualties. Such a weapon would have tremendous reach; exploded over the Pacific, EMP could be effective as far as the Mississippi River.[20] This, again, would have the major drawback of turning a regional war into a global one. However, smaller nonnuclear weapons have been developed that mimic EMP effects using microwave technology, but at least according to open literature, they are single-target systems without area-wide effects.[21] The PRC is said to be developing such weapons for use against U.S. aircraft carriers operating in the vicinity of Taiwan.[22]

Other types of attacks that could be confined to military units such as warships approaching Taiwan or in position off the PRC coastline in the hope that a war without collateral civilian damage might lessen global support for U.S. efforts and limit the impact of the conflict (at least in the short term) on the PRC's export economy.

Table 6-1 provides a matrix of the possible conflict initiations and some associated anti-access actions.

All the above are calculations that CCP leaders must make in contemplating the use of force concerning the Taiwan issue—along with the practical challenge of defeating the Republic of China (Taiwan) forces and occupying the island. But the most likely calculation concerns the ability of U.S. forces to break the overall PRC anti-access network and operate in the region of the Taiwan Strait.

Counter–Anti-Access Actions

In response to a Taiwan scenario, U.S. and allied/coalition forces (assuming they decide to intervene) also have the critical strategic choices of attempting to penetrate the PRC anti-access network, specifically in the Taiwan Strait area, or to break the PRC anti-access overall. The latter would seem the most practical choice because the anti-access network is probably well integrated, and assets can be shifted to cover the Taiwan Strait area. Also, attacks on the entirety of the network might force the PLA to decrease or suspend operations directed at Taiwan in order to respond. This might prompt a region-wide response from the PLA, but such a response (or previous preemption) could occur no matter the nature of U.S./allied/coalition actions.

Political, diplomatic, and economic activities would be obvious first steps. Economic pressures might have significant effects, given the export-dependent nature of the PRC economy. Seizure of PRC overseas assets, embargo of all imports from the PRC or subsidiaries located elsewhere, and declaration of all U.S. Treasury bonds in mainland Chinese hands to be invalid are but three of the significant economic measures

TABLE 6-1 Conflict Initiations and Anti-Access Actions

Initiation Paths	Alternative Characteristics	Sample Anti-Access Actions
(a) Bolt-from-the-Blue/ Preemption	(a1) Limited Preemptive	• Naval mining in the Taiwan Strait area • Ballistic and cruise missile strikes against military targets on Taiwan • EMP attacks on sensors and systems located in the Taiwan Strait area • Ballistic, cruise, aircraft, surface, and/or submarine attacks on warships operating near the Taiwan Strait area (with or without the declaration of a maritime exclusion zone • Surface-to-air and fighter aircraft attacks on aircraft operating near the Taiwan Strait area • Cyberattacks on U.S. military information systems
	(a2) Regionally Preemptive	• Ballistic, cruise missile, and/or bomber strikes against U.S. bases in western Pacific • Destruction of surveillance and support satellites focus in Pacific region • Attacks on U.S. and allied warships operating within the two island chains • Surface-to-air and fighter aircraft attacks on aircraft operating within the two island chains, including in Japanese and South Korean airspace • Antisubmarine and naval mine warfare throughout the western Pacific • Cyberattacks on U.S. military and civilian infrastructure information systems

	(a3) Strikes against CONUS ports (sea and air) of embarkation	• Long-range conventional ballistic missiles • Mining of U.S. ports by submarine • Destruction of U.S. satellites • Cyberattacks on U.S. military and civilian infrastructure information systems
(b) Gradual Escalation	(b1) Offensive anti-access tools	An increasing sequence of: • Diplomatic and economic efforts to forestall support for Taiwan • Diplomatic and economic efforts to undermine support for U.S. response • Actions described for (a1) limited preemptive attack • Actions described for (a2) regionally preemptive attack • Actions described for (a3) strikes against CONUS ports
	(b2) Relatively passive anti-access tools	• Diplomatic and economic efforts to forestall support for Taiwan • Diplomatic and economic efforts to undermine support for U.S. response • Naval mining in the Taiwan Strait area • Maritime blockade around Taiwan • EMP attacks on sensors and systems located in the Taiwan Strait area • Cyberattacks on U.S. military information systems

that could be applied. A global boycott of the PRC could be envisioned, although there would undoubtedly be violators. These actions would likely be resisted by members of the U.S. (and allied/coalition) business communities with assets in mainland China or dependent on Chinese imports, but such dissonance is natural in a democratic society.

Whatever the scope of the military response, U.S. actions in the Taiwan Strait region would follow a cross-domain sequence such as:

1. Degrade PRC sensors in order to blind the opponent to U.S./allied/coalition (and Taiwanese) actions. This could include cyberattacks, destruction of space-based systems, and electronic warfare attacks on PLA forces.
2. Maximum use of deception in U.S./allied/coalition operations.
3. Establish ballistic missile defense over Taiwan using sea-based platforms.
4. Establish maritime and air peripheral control around Taiwan through the use of submarines, long-range and tactical aviation, and carrier aviation as appropriate.
5. Reinforce Taiwan defenses. Taiwan must focus on its own anti-access effort, with similarities to the Battle of Britain.
6. Conduct sea-denial operations throughout PRC littoral waters, including naval mining, and submarine, surface, and air attacks, to prevent the transit or operation of any PLA naval or amphibious forces.
7. Conduct suppression of PRC air defenses (SEAD) in the PRC littoral regions, or as necessary.
8. Conduct strikes against all offensive PLA forces that might be engaged against Taiwan or U.S./allied/coalition forces (such as ballistic missile launchers).
9. Conduct air denial over PRC littoral territorial airspace in order to attrite PLA air assets or force them to operate away from the Taiwan Strait area.
10. Conduct strikes on PRC ports of embarkation to deny transit to amphibious or land forces targeted at Taiwan.
11. Achieve air superiority over Taiwan using land-based and carrier aviation.
12. Conduct sea basing on the periphery of the Taiwan Strait area in order to position ground forces to move onto Taiwan as needed.

Many of these operations would be conducted simultaneously in order to achieve the cross-domain synergy that is the focus of U.S. joint doctrine.

Sea Denial as a Counter–Anti-Access Technique

The particular geographic circumstances of a Taiwan conflict scenario is that if Taiwanese forces could survive the initial PLA ballistic missile and air attacks, the U.S./allied/coalition force may *not* have to achieve traditional sea control. Taiwan is 112 miles (180 kilometers) from mainland China, a short distance for missile combat but a long distance for amphibious craft to transit against opposition. It may be sufficient for U.S./allied/coalition forces to conduct their own sea- and area-denial campaign to prevent a physical crossing of PLA forces. This would be, in effect, a smaller "anti-access" operation but one conducted by the power-projection forces of an out-of-area power (therefore not qualifying as an anti-access strategy as previously defined). Naval forces conducting forward presence prior to (or during) a crisis would be the primary tools of this campaign, along with tactical aircraft and air defense based on Taiwan or at sea. Sea-based and land-based theater ballistic missile defense would also be components, albeit the PLAN would likely concentrate its missile attacks against the launch sites.

As discussed, sea denial has been considered an approach for strategically inferior navies and a component of regional anti-access. It is a posture that U.S. Navy (and U.S. Air Force) leaders may be uncomfortable in discussing, at least in theory. There has always been a general concern that the substantially larger resources required to build and maintain a global sea-control navy could be reduced by a presidential administration or Congress under the logic that sea denial is sufficient in itself to provide for a maritime defense. That is not true for a global trading economy like that of the United States. Economic security and sea power are always entwined. However, defense cost savings are always desirable objectives, resources are always limited to some degree, and the size of naval and joint forces is always a political question in a democratic state. However, the reality is that a global sea-control navy has within itself the capabilities of conducting a very robust sea-denial campaign. And the combat tactics of sea control are largely the same as those of sea denial. In both cases, Capt. Wayne Hughes, USN (Ret.), one of the most noted experts in naval tactics, advises that once hostilities start, the tactical objective is to "attack effectively first."[23]

A sea-denial campaign in the Taiwan Strait area may require the U.S. Navy to spend additional resources on a capability in which it has traditionally been reluctant to invest: naval mines. Since the days of the Royal Navy's global dominance, naval mines have been primarily considered the bane of sea power, a weapon to be used to hazard the great sea-control navies. As a U.S. Navy staff officer responsible for mine warfare once facetiously advised a conference audience, "mine warfare is the strategy of pusillanimous navies or nations with no navy." International law concerning mine warfare is largely misunderstood and sometimes confused with the international legal campaign against land antipersonnel mines. However, when sown near PRC ports of embarkation, modern mines and other autonomous undersea weapons might prove effective against the transit of surface warships and, possibly, submarines. At the same time, U.S. nuclear-powered attack submarines—the ultimate sea-control weapons—could hazard all PLAN maritime operations in the overall Strait area. A corresponding air defense/denial campaign, along with other cross-domain operations, could put PLA objectives at risk even prior to the arrival of the bulk of U.S. power-projection forces.

In this scenario response, certain anti-access tactics would be used in support of an overall counter–anti-access campaign. At the same time, Taiwan must develop its own anti-access network.

Attrition versus Precision

Advocates of precision guided munitions (PGMs) have long argued that precise targeting capabilities avoid the need for the area-wide attacks that were the standard for military operations in most major twentieth-century wars. From this perception, precision appears the opposite of attrition, now a term of opprobrium in military strategy. PGMs have certainly proven themselves in the wars since Desert Storm, and—as noted—their effective application by the U.S. military in force-on-force engagements was a primary factor in convincing potential belligerents to adopt an anti-access posture. However, anti-access and counter–anti-access warfare in such a scenario as presented previously would be a combination of combat techniques and weapon systems best described as "attrition warfare conducted with precision weapons."

A major factor is the huge number of short- and intermediate-range ballistic missiles in the PLA inventory, along with the density of the air defense systems involved. But the very nature of anti-access warfare provides incentive for saturation attacks, redundant targeting, multiple salvos, and redundant defenses. Damage effects on anti-access networks can have cumulative effects in the way that gradual chipping at the base can bring down a wall. Therefore, the objective of the anti-access network is to ensure the destruction of the counter–anti-access forces before they can get into position to commence their attacks on the network. Conversely, since multiple blows can exhaust the weapon inventory and personnel readiness of the anti-access defense, the counter–anti-access force has the incentive to maintain continuous pressure through massive attacks. Multiple simultaneous salvos can overload the information-processing or -transmission capabilities of sensors and combat systems or simply present more targets than any weapon system can engage.

With these incentives, the model may not be the one-shot, one-kill engagement that represents the growing doctrine of high-technology warfare. With high-technology systems on both sides and an ongoing struggle among stealth, deception, and detection, the model for combat may be the artillery engagements or antikamikaze "walls of steel" used in World War II. Precision (and nonprecision) weapons will be used in vast quantities as each side attempts to attrite the opponent's forces and break their will to continue.

Assessing an East Asia Campaign: Fundamentals

As in the previous examples, it is useful to assess the East Asia scenario in terms of the fundamental and supplemental factors we have applied to the historical cases.

Despite economic developments that might place the PRC in the lead in terms of gross national product, the United States remains *strategically superior* (1), a circumstance that is likely to continue at least to 2050. It is not a question of military technology or overall military strength, but that the United States had developed the logistical capabilities and alliance system to be able to project power to any region in the globe. This power-projection network was developed for over sixty years after a hot war victory and resulted in a cold war victory. In contrast, the PRC cannot project conventional military power beyond its region and can only threaten the United States with a nuclear strike that would result in devastating retaliation. Unlike the Soviet Union, the PRC does not have the nuclear arsenal that could allow it to imagine conducting a decapitating counterforce first strike. In terms of a regional conflict, U.S. global responsibilities could result in diffusion of effort and an inability to concentrate its total force, but it is unlikely that all its allies would be cowed into denying critical support. The PRC cannot send its forces across the Pacific; the United States can and has.

The major *geographic characteristics* (2) supporting a PRC anti-access defense against a U.S. advance is both the wide expansion of ocean that follow-up U.S. forces need to travel and the reality that Taiwan, the central focus of PRC efforts, is but a hundred miles away from mainland China. The United States can transfer aircraft and move troops by air but not heavy equipment. Under all but the slowest conflict development, forward-deployed surface warships would have to operate as the joint heavy artillery and land-support platforms. Cruise missiles and air strikes would be the dominant combat forces, with close coordination/synergy in both the other domains—space and cyber. Warships would also need to function as the theater ballistic missile defense (TBMD), supplementing any that Taiwan has on land. Obviously, the maritime domain (which includes the air above) would be the *predominant medium* (3) for combat.

The *criticality of intelligence and information* (4) and the use of deception *cannot* be overemphasized. The nature of the maritime geography presents a broad area of maneuver space. The reach of naval weapons would determine the extent that the maneuver space could be utilized to support Taiwan. This would pose some limitations, but targeting maneuvering forces would require the anti-access network to retain access to long-range sensors and surveillance systems. They are the most vulnerable parts of the network, and it is inherently crucial that the counter–anti-access forces destroy them. Another neutralization option is to use electronic warfare and deception to ensure that they provide false information to the anti-access decision nodes. Cyberwarfare would also be applied.

During the later stages of the Cold War, the U.S. Navy routinely deployed deception units with its operating forces. But in the immediate post–Cold War era of the brief new world order, this practice was reduced. It is logical to increase resources on deception capabilities and apply the full panoply of cutting-edge modern technologies to this effort.

The potential impact of *extrinsic events* (5) also *cannot* be overemphasized. It is here that authoritarian power holds advantages over democratic states as it uses the full range of what would otherwise be considered nongovernmental tools to affect events elsewhere. Even with its global economic activities, the PRC remains a regional power without extensive alliances that it is pledged to protect. It does have to consider the effects of its decision-making on its export economy, but it need not take into consideration the views or reactions of allies and partner states—Europe for example—that might consider the status of Taiwan to be peripheral to their own interests or overall global stability. The United States has political and economic interests in many regions, some of which frequently conflict. In addition, it has it own economic health to consider. A prolonged conflict over the Taiwan Strait would drain resources on both sides.

The PRC could also take advantage of unrelated crises occurring elsewhere and attempt to conduct a simultaneous fait accompli. Many commentators have remarked that America's recent fixation of the Middle East has blinded it to disturbing developments elsewhere. The attention spans of even the most perceptive decision-makers are limited. Extrinsic events can strain both America's attention span and resources. Military forces might be fully engaged elsewhere or stretch very thin as an East Asian scenario unfolds.

Assessing an East Asia Campaign: Supplementary Factors

From a purely operational standpoint, *strikes against regional bases* (A) are almost imperative to ensure that PLA forces can swiftly overcome resistance from Taiwan without outside intervention. They would greatly magnify the potency of the overall anti-access effort. However, as previously discussed, in the longer-term political-diplomatic view such strikes could create enemies of nations that—in a scenario of gradually increasing tensions—might not choose to directly intervene in PRC actions against Taiwan. The PRC has long argued that its relationship to Taiwan is purely an "internal matter" and has passed domestic laws to justify its perspective. Strikes against regional bases would turn this "internal matter" into permanent regional antagonism.

Therefore, the use of strikes against regional bases as an anti-access tactic would hinge on the internal dynamics of CCP decision-making (as do all major political decisions in the PRC) and on an overall perception of U.S. capabilities and resolve. If the CCP calculation is that U.S. forces could interfere with a successful campaign to incorporate Taiwan and might react to its initial actions but does not have a commitment to fighting a regional war with global implications, a "surgical" strike against U.S. forces in regional bases might be attractive—particularly if it could be done without much collateral damage and many casualties. A factor here is whether the scenario unfolds as part of long-term planning or as a Falklands/Malvinas situation in which internal pressure on CCP control is becoming unbearable. In the later case, military considerations could outweigh political calculations.

The problem for the counter–anti-access force—in this case the United States with units stationed at regional bases or forward-deployed at sea—is that it *cannot* assume that it can determine CCP/PRC intentions. It cannot know for certain, no matter the genesis or speed of the scenario, whether regional bases would be struck. It does know, however, that the PLA *has* the capability to launch a determined attack on bases throughout the region and must either deter or survive such action. This again puts a premium on the ability to defeat the PRC anti-access network as a prime element of deterrence.

Also as previously discussed, there are many operational incentives to incorporate *preemption* (B) into a PRC anti-access campaign. The longer the prehostilities phase in a Taiwan crisis, the greater the opportunity for the United States to send forward-deployed forces into the region and place its regional forces on alert. A declared PRC naval blockade (or "embargo") of Taiwan would not prevent the United States from landing troops and equipment by air, just as it did not prevent air traffic during the Berlin Airlift (during a road and rail blockade) and between the Soviet Union and Cuba during the Cuban Missile Crisis. In taking a page out of recent U.S. playbooks and declaring a "no-fly zone" above Taiwan, the PRC would place itself in the position of having to take the first shot—possibly at a civilian airliner.

Instead of such complications, a "surgical" preemptive attack that catches all opponents unaware can both speed the intended operational result and greatly enhance the effectiveness of the anti-access network in the case of a follow-on response.

Earlier we have described the incorporation of emerging technologies into military operations as an evolution. Weapon systems expand in terms of range, precision, and lethality as the underlying technologies upon which they are based advance. *Technological innovation* (C) is a factor in all war, whether it takes the form of an anti-access campaign or not. The question is: what is the relative balance of innovation between the combatants?

The United States has the reputation of holding the edge in the development of qualitative improvements to military technology. This became particularly true following the collapse of the massive Soviet military-industrial complex. But even when new technologies are protected by secrecy or export policies, the diffusion of the underlying knowledge is difficult to bottle. It is logical to assume that in an anti-access vs. counter–anti-access struggle, technological advances could create an operational advantage for a particular combatant—but that this advantage would likely be marginal in comparison to tactical prowess and resolution. It is not that tactical prowess and resolution can always overcome technological advances, a mistaken belief that doomed Imperial Japan. It is that in the modern context both sides are, to some degree, likely to have access to similar technologies. In the overall struggle of attrition using precision weapons, it is inappropriate to assume that a qualitative edge will be the decisive factor.

On the other hand, the ability to achieve *cross-domain synergy* (D) could be the decisive factor. Cross-domain synergy is desirable under all combat conditions. The side that can best coordinate its space and cyberwarfare assets with its maritime, air, and land forces would have the advantageous position. The traditional rule of thumb that an attacking force needs superiority at a ratio of three to one over defenders no longer applies on the modern battlefield. Rather, an attacker who can coordinate operations in all domains/dimensions can use multiple axes of approach to multiply their effectiveness. If it cannot, a capable defender has the advantage. Cross-domain synergy or its lack will always be a factor in any anti-access vs. counter–anti-access conflict.

Concluding Observations on the Scenario

The PRC is an authoritarian state in which public debate, particularly debate among military professionals, can be tightly controlled. It is therefore difficult to ascertain whether or not the strategic issues that are allowed to appear in the press reflect the development of actual PLA war-fighting doctrine. But if it does, it is apparent that the PLA recognizes that—at least for the immediate future—it must fight from an anti-access posture (no matter the term used).

Discussing how the U.S., allies, and partners could penetrate this anti-access posture in the event of actual hostilities does not "demonize China," but it can focus public awareness of the potential for conflict and bolster deterrence of an unfortunate event. Not examining the anti-access nature of this potential could actually increase its likelihood. Breaking the anti-access wall would be difficult. Bad news does not get better when it is ignored.

Noting that technology may not be a decisive factor is not a recommendation to reduce defense research and development. It simply acknowledges that there is no one "silver bullet" that can neutralize a robust, redundant anti-access network. Technological advantages must be combined with tactical acumen, deception, and cross-domain synergy if counter–anti-access forces are to be successful. Achieving this combination requires dedicated effort and prioritization of resources.

In prioritizing these defense resources, it must be recognized that the high-intensity, attrition-with-precision-weapons combat of an East Asia anti-access scenario is the most difficult challenge that U.S. military forces could face in the near future.

Chapter 7

Southwest Asia: Asymmetrical Tactics and Economic Threats

A struggle between U.S. forces and those of the Islamic Republic of Iran fighting from an anti-access posture would have some similarities to an anti-access vs. counter-anti-access conflict in East Asia. But this Southwest Asia scenario would have even more differences. The bottom line is that Iran simply does not possess the resources of the PRC, and its anti-access network would be limited in comparison. Iran lacks the sustainability to continue a prolonged anti-access conflict against the United States. Given the current order of battle for U.S. power-projection forces and the Iranian anti-access network, under a sustained U.S. attack the Iranian network would crack in a relatively short amount of time. That is one of the reasons that Iran is pursuing the development of nuclear weapons. Used in a tactical role, nuclear weapons could be the ultimate anti-access system, obliterating regional bases, ports of embarkation, and—with reasonably accurate ISR—fleet concentrations at sea, particularly in the narrow confines of the Persian Gulf.[1] Many of the weapon systems presently incorporated into the anti-access network are the same or similar to those in the East Asia scenario because Iran receives most of its weaponry from China, Russia, and through joint development with North Korea. Not all these systems are the very latest models, and many may not be as capable as those operated by the supplying nation. However, despite these similarities, the potential causes for conflict, the objectives, the nature of extrinsic events, and the overall sequence of operations would differ. Iranian anti-access operations would likely include considerable use of asymmetrical warfare tactics—including strikes against regional bases or ports of embarkation by proxy forces and terrorist groups, and there could be greater incentive for the use of weapons of mass destruction (particularly nuclear weapons if they can be developed). U.S. forces in the operation should assume that they might be under chemical weapon attack.

Details of the overall Iranian defense posture are available from a number of independent sources, such as IHS Jane's (formerly Jane's Defence Group), the International Institute for Strategic Studies (IISS), and the Stockholm International Peace Research Institute, and official sources such as the DOD's *Annual Report on Military Power of Iran* (prepared for Congress).[2] These details, such as order of battle, will not be replicated here. Rather, the focus will remain on the principles of anti-access as they apply to Southwest Asia.

Differences in the Scenarios

As a first consideration, initial geographic distances between the contending forces would be much shorter than in the East Asia scenario. The northern section of the Persian Gulf averages 130 nautical miles (149.5 miles/234 kilometers), but narrows to 70 nautical miles (80.5 miles/126 kilometers) from Dubai to the Iranian coast, and 21 nautical miles (24 miles/38 kilometers) at the narrowest part of the Strait of Hormuz. Flight times for U.S. aircraft flying from bases in the Arab states of the Persian Gulf would be relatively short, although the distance between Kuwait and Iran's capital Tehran is 486 miles (783 kilometers). The U.S. Navy utilizes Bahrain as a major logistical and support center, routinely keeping warships—sometimes including a carrier battle group—deployed within the Gulf. U.S. air and ground forces have access to a suitable number of bases within the region. All in all, the United States regularly maintains substantial military forces a relatively short distance from Iranian territory. This would allow for considerable striking power against the Iranian anti-access network.

This does not mean the United States can be sanguine about its prospects in facing an Iranian anti-access strategy. Retired U.S. Air Force colonel Mark Gunzinger of CSBA, author of the most extensive public report on Iran's anti-access potential, maintains that "assumptions [that] served the U.S. military well for the past thirty years . . . should not be assumed [to be] suitable for ops against capable A2/AD complexes."[3] Gunzinger identifies these "traditional" assumptions as:

- Close-in bases would be available for early-arriving fighters and command-and-control elements.
- Naval units could operate within range of their target areas.
- The United States would have a near monopoly on precision guided munitions.
- There would be no real threats to refueling and logistical operations.
- U.S. C^4ISR networks would remain secure.[4]

Although Gunzinger is quite correct in identifying these assumptions as features of U.S. joint planning for post–Cold War contingencies, they have not necessarily been the traditional assumptions of naval war-fighting strategy. The U.S. Navy and Marine Corps have certainly benefited from being able to operate closer to shore and have adopted a littoral focus. And the ranges of many current naval weapons could be extended by a variety of operational techniques. It must be conceded that—overall—U.S. forces have been "spoiled" into assuming that they will operate at close range and that their bases will have relative sanctuary from attack. This has actually been a reasonable assumption given the contingencies that they have faced in recent years, with an emphasis on counterinsurgency, humanitarian assistance, and conflicts against an Iraq that did not apply anti-access techniques. It is not, however, appropriate to an anti-access environment.

Another consideration is the potential of Iran to focus on asymmetrical warfare techniques such as extrinsic attacks and the use of terrorism within the territory of U.S. allies and partners, or even the United States itself. The primary extrinsic attack would be against Israel in order to turn any conflict with the West into an Islamic jihad. This might not sway the governments of the Sunni Arab states from denying the United States base access, but it could create unrest within them. Additionally, Israel would likely conduct a counterattack, further inflaming jihadist sentiment. This, of course, might backfire if Israel and Iran trade devastating attacks. Western forces might have to enter both countries under the auspices of the United Nations in order to provide security and relief, and possibly a democratic (or at least pseudo-democratic) government in Iran.[5]

Terrorist attacks in Western nations remain a possibility, although organized Iranian proxies such as Hezbollah are largely confined to Lebanon and adjoining areas.[6] It is hard to tell the true reach of Iranian terrorist networks, but since 1984 the State Department has designated Iran a state that directly sponsors terrorism. Most Islamic terrorist acts have been carried out by extremist Sunnis, and it is unknown how much Shiite calls for jihad would influence them.[7] In any event, attempts to direct terrorist attacks against regional bases and ports of embarkation should definitely be considered part of the Iranian anti-access strategy.[8] Terrorist attacks against general civilian targets could also be used to generate the extrinsic events that could distract or persuade the United States, allies, and partners from continuing counter–anti-access operations. It is unclear if the Iranian theocratic leadership—so used to denouncing the dissolution of Western society—understands the Pearl Harbor syndrome.[9] Initial U.S. actions in Afghanistan after 9/11 might be instructive to them in that regard.

A third consideration—as previously mentioned—is the greater incentive for the use of nuclear weapons or other WMDs as part of the anti-access strategy. Given the imbalance in conventional weapons, Iran's leaders appear to be adhering to General Sundjari's edict that one should not confront the United States without possessing nuclear weapons. That does not mean that they would use a future nuke in an initial anti-access operation. As with all nuclear weapons, the power of deterrence lies in the potential for use rather than actual use, with nuclear blackmail of regional U.S. partners being the most productive objective. From the Iranian perspective, nuclear weapons could also dissuade the United States from participating in a regional conflict because of the potential for high levels of casualties. In contrast, actual nuclear use could provoke a nuclear response. Yet the potential for use as a later part of a failing anti-access campaign against a counter–anti-access force set on regime change cannot be discounted. Despite Kenneth Waltz's deliberately provocative suggestion that an Iranian bomb would ultimately lead to a more stable Middle East peace, there can be no certainty that a messianic, potentially jihadist theocratic regime would refrain from nuclear use in the same way as the hostile superpowers in the Cold War.[10] Wars of religion do not necessarily follow Clausewitzian logic.[11]

Another distinction that may have an effect on anti-access strategy is that Iran essentially has two rival militaries with bifurcated operational responsibilities. The Armed Forces of the Islamic Republic of Iran (IRIA) has control of the "traditional" armed forces and is charged with protecting the territorial integrity of the state. The Army of the Guardians of the Revolution—known in the West as the Iranian Revolutionary Guard Corps (IRGC)—is tasked with the "defense of Islam" and "preservation of the revolution," which translates as the preservation of the regime. The IRGC has control over the national missile force, internal security, and the training of proxy foreign militaries; specializes in asymmetrical warfare operations; and is reported to have responsibility for operations in the Strait of Hormuz. On paper the IRGC is part of the IRIA, but in reality it operates separately and has much greater political and economic power and influence. The IRGC commander, not the IRIA commander, sits on the supreme national security councils, and IRGC veterans dominate political positions. The IRGC is also responsible for the national defense industry and runs many domestic businesses. The bifurcated defense responsibilities make it unclear as to how well all the elements of an Iranian anti-access network would be coordinated in event of actual operations. A peculiarity to the Southwest Asia scenario, it might prove to be a weakness in the outer wall.

In his study, Gunzinger refers to the unique characteristics of the Iranian anti-access strategy as "A2/AD with Persian Characteristics," which he describes as "an asymmetrical 'hybrid' A2/AD strategy that mixes advanced technology with guerrilla tactics to deny U.S. forces basing access and maritime freedom of maneuver."[12] His Persian Characteristics include those discussed previously along with the use of less technologically sophisticated systems in innovative asymmetrical tactics.

One such tactic that is often suggested by other commentators (and discussed by Gunzinger but without claiming that it would be particularly effective) is the use of relatively lightly armed fast attack craft in "swarming" attacks on individual warships. This tactic is similar to attacks conducted in the Iran-Iraq War, in which the IRGC used a fleet of high-speed patrol craft, manufactured by the Swedish company

Boghammar Marin AB and armed with short-range rocket-propelled grenades, to attack larger vessels. Since then, the term *Boghammar* has been applied to a wide range of small, improvised naval vessels. More recently envisioned is the use of multiple waves of small craft as suicide weapons, or in the form of explosive-packed remote-controlled unmanned surface vehicles (USVs), like a deadly swarm of insects. The tactic has captured the imagination of many analysts since it is presumed swarms of such vessels can be constructed at a relatively low cost—functioning as multiple Davids against a stronger, but slower, individual Goliath.[13]

In contrasting a likely Iranian anti-access strategy—endowed with modest resources—with that of the PRC, Gunzinger points out that the latter would have to cover much greater distances of maritime space in order to defend against an attack in the Pacific. Iran's network can instead focus on the narrow confines of the Persian Gulf/Strait of Hormuz region.[14] Earlier we noted the shorter distances as giving greater advantages (than in the Pacific) to land-based tactical aviation operating in a counter–anti-access mode. It also must be admitted that the narrower confines do potentially heighten the effectiveness of the less-dense Iranian arsenal of diesel-electric submarines, naval craft, and ballistic and cruise missiles, along with the much shorter range but denser inventory of guided rockets, artillery, mortars, and missiles—G-RAMM.

It should also be noted that Iran also has a modest inventory of medium-range ballistic missiles (MRBMs) capable of striking as far as Eastern Europe (obviously Israel as well). Most were developed with North Korean assistance but are under indigenous production. However, the actual military effectiveness of the MRBM force is unknown and suspect.[15] They would probably be used as terror weapons against population centers of opponents. In any event, it is a valid observation that although Iranian anti-access capabilities—particularly in reconnaissance, detection, and surveillance—are much less robust than those in East Asia, they also have much less space to cover and, by design or default, are optimized for the narrower confines of the regional geographic conditions.

Causes of a Conflict

Throughout its existence, the Islamic Republic of Iran has expressed its hostility toward the United States because it supported Shah Mohammed Reza Pahlavi, because it supports the Arab states of the Persian Gulf region, and of course because it supports the existence of Israel. This continues despite the fact that the U.S.-led removal of Saddam Hussein provided Iran with a strategic windfall. Its most determined enemy and buffer to expansion of its regional influence was thus removed. Additionally, U.S. forces removed the Taliban from power in Afghanistan, another hated enemy who had killed Iranian diplomats.

The hostility also stems from the fact that the theocratic autocrats of Iran view the United States as an enemy of the political influence of Shia Islam and is thus the "Great Satan" that threatens the natural course of global Islamic conversion.[16] A more worldly and immediate political conflict lies in the opposition of the United States and many other nations to Iranian development of nuclear weapons. The Iranian government views its development of weapons of mass destruction as a right of its statehood and a necessity for protecting its integrity and continuation of the theocratic regime.[17] Most members of the United Nations are concerned with this step in global proliferation, as well as a perception that the Iranian government would willingly undertake offensive nuclear actions, such as an attack on Israel—a perception fueled by the messianic rhetoric of the theocratic regime. Since the United States has taken the lead in developing international sanctions against Iranian nuclear weapons and may or may not be responsible for cyberattacks on the Iranian weapon development infrastructure, Iranian government hostility naturally continues.[18]

The Iranian government also makes clear its general hostility to the Sunni Arab regimes on the western side of the Persian Gulf and its belief that it should have decisive influence over matters in the Gulf region. Like most of the Gulf states, Iran is economically dependent on oil exports, with oil constituting 76 percent of export earnings and 62 percent of government revenues.[19] Like that of the other Gulf states, most of Iran's oil travels by tanker through the Strait of Hormuz, although Iran does have coastline outside the Strait bordering the northern Arabian Sea. With sanctions imposed on its oil exports as a consequence of its continued nuclear weapon efforts, the Iranian government has periodically threatened to close the Strait if the United States or other nations take other actions against it. As a practical matter, the PRC is Iran's largest customer and an opponent of sanctions; the closure of the Strait would therefore harm Iranian friends as well as foes. This fact has led many commentators to suggest that the Iranian threat to the Strait is overblown.[20]

Unlike in the East Asia scenario, there are no irredentist claims that motivate Iran toward hostility to the western Gulf states. Arab Shiites are not ethnic Persians, and the Iranian posture as the protector of Shia Islam does not necessarily translate into a controlling position in the Shiite communities in the Arab states (as opposed to being a useful source of support).[21] The cause of a conflict in the Hormuz Strait region is less likely to be territorially based but rather would involve an attempt to demonstrate regional power or be owing to a threat to the survival of the theocratic regime. With the exception of the nuclear weapon issue, it would seem that outside states have little incentive to directly involve themselves in removal of the regime—and few practical means to achieve regime change. But the Iranian government's links with groups that use terrorism as a political tactic do place it within the crosshairs of affected nations and made it a member of President George W. Bush's controversial "axis of evil."

Reports have routinely indicated simmering popular dissatisfaction with the theocratic regime, which some characterized in 2009 as a "green" or "velvet" revolution.[22] Despite enthusiastic reports by Western media and the U.S. government's wishful view that it would herald regime change in Iran, it became clear that the discontent did not necessarily represent a Western-inspired movement. However, the Iranian government has frequently made it even clearer that it views all internal opposition to the regime to be the work of Western states, notably the United States or Israel. In the case of a significant challenge to its authority, it is possible that the theocratic regime might make good on its threat to close the Strait—not as a means of inciting nationalist fervor—but as a means of lashing out against the perceived supporters of its internal opponents. This list of supporters could also include the Arab emirates and Saudi Arabia.

Possible Iranian actions that could elicit a U.S./coalition military response include:

- An Iranian attack on the Gulf emirates or Saudi Arabia
- Direct Iranian support for Shiite rebellions in the Gulf emirates or Saudi Arabia
- Iranian military intervention in Iraq
- Long-range or proxy attack on Israel
- Closure of the Strait of Hormuz in support of a military or political objective

Any initial U.S. response would include the neutralization of the Iranian anti-access network.

Shifting the Nuclear Weapon Paradigm

Tactical nuclear weapon development is meant to compliment conventional aspects of the anti-access network. Some of the implications of the Iranian quest for nuclear weapons have been previously discussed, but it is important to recognize that the possible inclusion of nuclear use as a part of anti-access strategies—a potential feature of the Southwest Asia scenario—would mark a profound shift in the nuclear weapon paradigm that has been the basis of international arms control and nuclear nonproliferation efforts.

For sixty years American (and other) strategists have assumed that the ultimate purpose of nuclear weapons is strategic nuclear deterrence: they deter the use of corresponding nuclear weapons by an opponent and ensure a general condition of peace in an environment of ideological hostility. This is the premise of Waltz's suggestion that an Iranian bomb would deter Israel from using its own probable nuclear weapons and therefore could somehow contribute to Middle Eastern stability. But nuclear weapons as elements of anti-access introduces the premise that their use against conventional attack, or integrated with a conventional attack on a nonnuclear opponent, is a logical element of modern warfare. Many analysts have identified nuclear use as a possible aspect of asymmetrical warfare, and the relationship between the asymmetrical warfare concept and anti-access has been previously discussed in this book.[23] This has the potential for making preemptive nuclear attacks in a regional war more likely than they have been in the past. This can also be the basis for a renewed global nuclear arms race.[24] It has certainly spurred global interest in TBMD, itself a feature of any anti-access vs. counter–anti-access conflict.

Ironically enough, Sundjari's premise that nuclear weapons should be sought prior to any confrontation with the United States is much less valid than it might seem at first glance. The United States maintains a strategic nuclear arsenal that would dwarf that of any of the potential anti-access states, with the exception of Russia. Nuclear weapons poised against U.S. allies or partners could have a deterrent effect on their support for U.S. actions or could even deter U.S. actions for fear of harm to an ally. But no nation that contemplates tactical nuclear use against U.S. military forces can rule out the possibility of a response in kind. From that perspective, regional neighbors would have much more to fear from nuclear weapon development than does the United States.

In any event, a situation where the same tactical ballistic missiles could have either nuclear or conventional warheads presents a dilemma for the premise of nuclear deterrence and stability. And it is a situation toward which the Southwest Asia scenario appears to be moving. It would admittedly make a counter–anti-access campaign more operationally difficult, but it would also mark a shift in assumptions concerning the likelihood of nuclear war, a likelihood that was so greatly reduced following the end of the Cold War. This time the question of what constitutes stability might be too complex to leave to professors.

Geography and World Economy

The Persian Gulf with its Strait of Hormuz is the defining geographic condition of the scenario. It would be more difficult for U.S. power-projection forces to strike at Iran from axes other than across the Gulf. And striking across the Gulf—even with a significant number of strikes by land-based tactical aviation—requires the transportation of heavy equipment and naval capabilities by sea. With the exception of nuclear weapons, the potential of Iranian blockage of the Strait of Hormuz could be its most effective anti-access tool—if it could be sustained. That is why U.S. naval mine countermeasures (MCM) are critical for any counter–anti-access campaign. Despite traditional neglect, the need for new MCM have been recognized, and the U.S. Navy, along with operational parties, has increased its MCM force and conducted a number of MCM exercises in the wake of the latest round of Iranian threats.[25]

Yet threats to close the Strait of Hormuz are in themselves useful tools in the simmering confrontation with the Western world because of the effect they can have on global economic fears and the price of oil. While oil shortages that would ensue from an actual closure might be alleviated by the release of oil reserves and coordination among oil-importing states, oil prices routinely spike when the Strait appears genuinely threatened, subsiding only when it is clear that the United States and its partners have sufficient forces in the Gulf region to ensure that any closure would only be temporary.[26]

It is also important to remember that Iran did conduct a mining campaign in the international waters of the Strait region during the Iran-Iraq War in an effort to reduce tanker traffic and the trade flow to Iraq. When a mine struck the reflagged oil tanker *Bridgeton* in 1987, then being escorted by U.S. Navy warships, the United States greatly increased its MCM capabilities and its overall naval presence throughout the Gulf. At that point, U.S. warships did not strike mines and had been damaged by presumably accidental Iraqi air attacks, somewhat muting the impetus to retaliate directly against Iran. But in 1988, USS *Samuel B. Roberts* did hit a mine, prompting a retaliatory strike against Iranian oil platforms and naval units. Iran did not opt for further confrontation, and a few months later (following the accidental downing of an Iranian jetliner by USS *Vincennes*) agreed to a U.N.-sponsored ceasefire with Iraq.[27]

The point of recalling this is threefold. First, any anti-access vs. counter–anti-access confrontation with Iran would likely involve some form of restriction of the flow of oil through the Strait of Hormuz, which would in turn have an effect on the overall global economy.[28] The economic impact could function as an extrinsic event that influences U.S. decision-making, particularly if pressure is applied by the PRC and others to "suspend hostilities" in a way that leaves Iranian objectives achieved. Second, in any prehostilities phase, a threat to close the Strait could also prove a useful way of introducing extrinsic events—reactions from the world economy—without actual conflict. Finally, threats and/or actual attempts to close the Strait could conceivably have an effect opposite to strengthening Iran's anti-access posture. It might instead strengthen the resolve of the United States and partners to break the entirety of the

Iranian anti-access network in order to eliminate the threat completely. Whatever the outcome, the potential scenario continues to demonstrate the predominance of geography in an anti-access vs. counter–anti-access confrontation.

Counter–Anti-Access Activities prior to Conflict

In reality, the United States and its partners can be seen as already conducting nonmilitary counter–anti-access activities by imposing sanctions against Iranian nuclear weapon development. At the same time, Israel, another state, and/or perhaps a nongovernmental hacker community conducted cyberattacks against the Iranian nuclear development infrastructure using the Stuxnet computer worm.[29] Covert actions intended to impede the Iranian nuclear weapon development process, along with operations against terrorist activities sponsored or encouraged by the Iranian government, have been referred to by journalists and commentators as "the secret war with Iran."[30]

From a counter–anti-access perspective, the purposes of such political, diplomatic, economic, and covert activities conducted prior to a potential conflict are (or should be) to stem the expansion of the anti-access network, eliminate supporting intelligence and terrorist activities, and reduce the impact of extrinsic events that could be fostered by the anti-access state. The forms of these activities would not necessarily differ from those taken against any hostile forces; however, the overall objective should include reducing the inventory of anti-access tools available for use in conflict.

Many presumably unrelated nonmilitary activities can play some role in deterring an anti-access conflict or shaping its features. As an example, arms embargoes and general efforts to reduce unregulated international weapon traffic can serve this purpose, although most significant anti-access weapon systems are transferred on a state-to-state basis. But significant weapons originating in Iran and North Korea intended for Hezbollah and other nonstate forces have been intercepted. Along with the potential for preventing their use in local conflicts, this could reduce the incentive for such forces to operate as proxies or adjuncts to an overall anti-access strategy.

Admittedly such a result is largely a by-product of actions motivated by general concerns about weapon trafficking and proliferation that have no direct connection to specific war-fighting strategies, anti-access or not. This very admission illustrates the benefit of analyzing the development of anti-access strategies from a much broader context than simply addressing it as an operational military problem.

In the case of the Southwest Asia scenario, the broader context includes the internal contradictions of a governing regime whose fundamental justification appears to contradict the perceived interests of most of the national population. Expert commentators with wide differences in political philosophy have identified the fact that governing principles and objectives established by Ayatollah Ruhollah Khomeini following the 1979 revolution appear to be in opposition to historical Iranian national interests and Shiite Islam as it had been traditionally practiced. One critic has described the evolving Islamic Republic as an effort to "Arabize" Iran and eliminate Iranian/Persian national consciousness.[31] Reminiscent of a "nuance" of the East Asia scenario, the dominant IRGC is formally pledged to defend Islam and the Khomeinist revolution, not the nation of Iran. Although detailed discussion of support for internal revolt is not a part of this book, intrinsic events would seem a significant possibility in eliminating the potential for an anti-access vs. counter–anti-access campaign in Southwest Asia, since the principles of participatory democracy seem accepted by the population, whose manipulation by the theocratic autocracy is relatively apparent and perhaps widely resented.[32]

Counter–Anti-Access Actions during Conflict

In his CSBA report and subsequent briefings, Gunzinger has mapped out recommended actions for U.S. and allied forces to take in order to reverse an Iranian closure of the Strait of Hormuz, degrade Iran's anti-access capabilities, and "impose costs" on aggression. The plan does not call for an actual invasion of Iran, but it does suggest three other "theater campaign lines of operation" that could also be alternative approaches or even objectives of a U.S. response: countering weapons of mass destruction, arming counter–proxy groups with G-RAMM, and using unconventional warfare to bring about regime change from within.[33]

WMDs, particularly nuclear weapons; attacks on U.S. forces or other out-of-region targets by proxy groups; and maintenance of tight control (a curtain, if you will) over Western or non-Islamic influences (as defined by the regime) on Iranian civil society can all be considered aspects of the Iranian anti-access network, although the latter certainly suggests a non-military response. In reality, such tight control does not exist today, and Iranians are not isolated from Western civil society, having contact with the large number of Iranians living in Europe and the United States. But the regime does have an infrastructure in place—through the IRGC—to attempt to impose greater control.

Gunzinger titled his study *Outside-In* to emphasize his view that Iranian anti-access systems, particularly ballistic missiles, have put all the regional bases to which U.S. forces could deploy at risk. His view is that *strikes against regional bases* (A) should be expected in any conflict with Iran. The opening phase of the anti-access campaign would therefore need to be conducted by forces "outside the reach of Iran's anti-access threats [in order] to gain the advantage."[34] The recommended campaign would consist of three conceptually sequential actions or phases (although overlapping and possible even simultaneous):

1. Dispersing and deploying U.S. forces (primarily air forces) to secure staging bases located outside the range of Iranian missiles
2. Blinding the enemy to "shrink the anti-access threat ring"
3. Creating "enabling pockets of air and maritime superiority . . . to create lodgments near [the] Strait."[35]

Initially outside-in operations would be conducted by long-range attack aircraft/bombers operating from secure bases and cruise missiles operating from warships in the Indian Ocean. Operations would be conducted from multiple axes. This would be supported by electronic warfare, cyberwarfare, and nonkinetic attacks in other domains. The attacks would focus on sensors and other means of targeting the counter–anti-access forces. As this blinding operation unfolds, attacks could be conducted against command-and-control and weapon system concentrations, particularly air, naval, and missile bases and air defense systems, using precision guided munitions. PGMs are reasonably assumed to ensure low civilian casualties, although there are no indications that the Iranian population, with nationalist sentiment, would consider that a sufficient justification to turn against the regime.[36]

Once the information-denial operations, command dissolution, and SEAD and destruction of coastal defense begin to take effect, "joint theater entry operations"—formerly known as amphibious assaults—would be conducted in areas where Iranian forces could continue to hazard the Strait, such as near-Strait littoral regions and Iranian-held islands in the Gulf.[37] The objective would be to force the Iranian regime to acknowledge that it does not have the power to block the Strait and to renounce such efforts in the future. With sole focus on counter–anti-access operations, Gunzinger does not attempt to define a detailed end state following the operations.

Assessing a Southwest Asia Campaign: Fundamentals

Assessing the Southwest Asia scenario in terms of the fundamental elements verifies the previous discussion of the scenario and its characterization as an anti-access conflict. One nuance needs to be acknowledged, however. An anti-access vs. counter–anti-access struggle would logically be conducted on multiple axes and combat would primarily consist of missile, countermissile, aircraft, and counteraircraft operations. It is more appropriate to consider the maritime environment (the Strait, the Gulf, and the Iranian littoral) as the *focus* of the operation rather than the *predominant medium* (3) for combat operations. In this case, the geography of the region demonstrates its defining characteristics in the narrowness of the water space. Naval forces would be significant elements throughout operations and would be primary in conducting the objective of securing the Strait. But the air over land would be the initial, and likely most difficult, battleground until the Iranian anti-access network is suppressed. Ultimately, the struggle—assuming the cause is closure of the Strait—would revolve around the maritime environment, and there is no argument that the analytical model would be invalid.

Clearly the United States is and will remain *strategically superior* (1) to Iran in terms of political, economic, and diplomatic strength, military technology, and overall military strength, and continue to have the logistical capabilities and alliance system to project power to any region in the globe. In contrast, Iran cannot project conventional military power beyond its region and can only directly threaten the United States through terrorist attack. It is even less likely that Iran could cow

U.S. partners into denying critical support. Iran recently sent two warships into the Mediterranean with much publicity, but that was a strategically insignificant act for a navy that cannot sustain itself.[38] It is the imbalance of military power that drives Iran to an anti-access strategy, including its desire for nuclear weapons. Whatever its rhetoric concerning the weaknesses of the West, the Iranian regime clearly recognizes the strategic superiority of the United States and a U.S.-led coalition, and seeks methods of neutralizing it within the Middle East region.

The *geographic characteristics* (2) of the Persian Gulf region and the *predominance* (3) of the maritime environment as the focus of operations have already been discussed. The status of the maritime domain of the Gulf (which includes the air above) would be the likely casus belli, and (with the exception of Iraq and Syria) it separates Iran from those states which it most greatly desires to influence or control. Iran can strike the Gulf states with missile attacks, but its troops would still have to cross water in order to exert control.

As in all anti-access warfare, the *criticality of intelligence and information* (4) and the use of deception must be emphasized. There is a much greater imbalance in this regard in comparison to the East Asia scenario, with the United States possessing considerable advantages. But it is conceivable that another state could share information with Iran that could bolster its awareness of the further reaches of the battle space. In any event, sensors and information systems remain the most vulnerable parts of the network, and it is always inherent that the counter–anti-access forces destroy them or feed them false or deceptive information. Cyberwarfare is an integral aspect of this effort.

The potential impact of *extrinsic events* (5) is difficult to judge. Iranian diplomacy would obviously seek to sunder the United States from its allies and potential partners, but a closure of the Strait would have global effects that would not seem to support the Iranian position. Many nations—including the Gulf states—appear to be publicly uncomfortable with the prospect of a U.S. or Israeli preemptive strike on Iran's nuclear development complex to prevent its acquisition of nuclear weapons, even if they are concerned about the results of such proliferation. But that does not mean they would countenance Iranian control over the Strait.

The use of terrorist attacks by IRGC agents or proxies in nations that might support U.S. forces should be expected. But such attacks are just more likely to galvanize support for action against the Iranian regime than suppress it, particularly if final U.S. objectives remain on par with those of the Gulf War.

A corresponding crisis in some other part of the world could conceivably take the focus off the Gulf region, but it would have to be a particularly catastrophic event, such as a war in the Taiwan Strait.

Assessing a Southwest Asia Campaign: Supplementary Factors

Strikes against regional bases (A) are expected elements in any Gulf region anti-access struggle. If the United States were not a factor, such strikes could be carried out against the Gulf states with relative impunity. They do not currently possess the capability to strike back, which is, of course, why they rely on U.S. support. Making regional bases untenable could delay U.S. forces, though not defeat them. But in absence of nuclear threats, it seems the only Iranian action that could come close to gaining the time for attrition and/or extrinsic events to have any effect.

Although there are operational incentives to incorporate *preemption* (B), a preemptive strike by Iran in the Gulf region would hold echoes of Pearl Harbor as a tactical surprise but a strategic failure. A preemptive attack would simply confirm the arguments that the Iranian regime constitutes a regional, even global danger and provide the justification for an extensive, perhaps regime-changing response. But as in a Taiwan crisis, the longer the prehostilities phase, the greater the opportunity for the U.S. to deploy forces into the region and place its regional forces on alert. Yet it could be that drawing greater U.S. forces into the region would be a key element of Iranian strategy, with a sudden strike on regional bases intended to be a morale-shattering blow that would curb popular support for American intervention. It would be unlikely to succeed, but again it is one of the few actions that provide time and allow attrition. *Technological innovation* (C) would not seem to be on the side of Iran under any circumstances, although possession of nuclear weapons could be seen as a technological innovation for the IRGC. Access to advanced military technology from other nations would have to occur prior to conflict, and U.N. sanctions are intended to prevent it. As porous as sanctions may sometimes be, the chance that a "war-winning" innovation would be transferred to Iran is very, very slim. Iran's most persistent military development partner is North Korea.

Iran can be expected to use existing weapons in asymmetrical ways—sidewise technologies, as they have been described. But success would require a pretty complacent opponent.

The ability to achieve *cross-domain synergy* (D) should be a great advantage for the counter–anti-access forces of the United States, allies, and partners.

Concluding Observations on the Scenario

The Southwest Asia scenario presents all the aspects of an anti-access vs. counter–anti-access campaign. It does have specific "Persian Characteristics" shaped by limited anti-access capabilities and much tighter geography for maneuver. Practical aspects of AirSea Battle are still a mystery, but it seems logical that the same emphasis on an air and naval response applies to the scenario. Like a potential conflict with the PRC, there is simply no cause for conflict with Iran that would seem to necessitate major land operations. Regime change from without—even if that were seriously considered as a U.S. policy—would be immensely more difficult than in Iraq, which had a much less stable regime.

As suggested by Gunzinger, some operations on land in the littorals by U.S. Marine expeditionary forces, special operations forces, and airborne troops might be necessary to neutralize short-range weapons threatening the Strait of Hormuz, but they would most likely be limited in scope. In a Strait of Hormuz closure situation, Western objectives would most likely mirror the first war against Iraq (liberation of Kuwait) rather than the second. It is a stretch (and perhaps even a bit facetious) to suggest that a participating coalition might even include PRC forces, but in terms of the economic value of oil flowing from the Gulf, closure of the Strait would have an even greater impact on the PRC (as well as Japan and South Korea) than on the United States.

A conflict over nuclear weapon development or initiated by a collapsing regime would present different problems, but it is difficult to see how a successful counter–anti-access campaign would need to be followed by major ground combat. Ground combat would face other geographic constraints, because Iran is essentially bisected by mountain ranges and does not present the almost limitless, flat desert terrain of Iraq. The population of Iran is over three times the size of Iraq and does not appear favorable to "liberation by outside forces."

Anti-access operations by Iran would largely be offensive in nature—consisting of missile attacks on regional bases and U.S.-supporting states, as well as naval forces that could be reached.[39] A closure of the Strait would essentially be a preemptive action, but it is difficult to see how the Iranian regime could think that it would retain control given the relative imbalance in military capabilities. It is unlikely that other states would accept the fait accompli and pressure the United States not to respond in the hope that a "deal could be reached." Terrorist attacks by proxies on the United States would embolden American actions rather than be discouraging extrinsic events. But as noted, the United States and its allies do not have a clear understanding of the actual decision-making process of the theocratic regime, and if motivation were to match the routine messianic and hostile rhetoric, a counter–anti-access campaign might need to become a reality.

Gunzinger recommends a "more balanced mix of short-range and long-range capabilities for surveillance and strike," which is a euphemism for stating that the U.S. Air Force and, to some extent, the U.S. Navy have invested too little in long-range bombers and attack aircraft and too much in shorter-range fighters (such as the F-22, now F/A-22), which are optimized for air-to-air combat from regional bases or near-shore carriers. That appears a most reasonable recommendation for increasing the effectiveness of U.S. counter–anti-access operations in the scenario. Complementing that, he suggests increasing the hardening of the bases existing in the region with the understanding that hardening against attack has limitations that do not make it a substitute for increasing the range of the overall counter–anti-access force.

Gunzinger also expresses concern about the falling numbers of submarine missile-launch tubes as the result of the decommissioning of the four converted nuclear-powered guided-missile submarines without replacements programmed for around 2025. He describes his recommendation for strengthening U.S. undersea-warfare/land-attack capability by modifying the current U.S. attack submarine design to increase its missile load as "almost a no brainer."[40] From a counter-anti-access warfare perspective, it is hard to disagree with his recommendation. The obvious requirement is the appropriate allocation of resources, always a struggle in the face of competing requirements.

The likelihood of the Southwest Asia anti-access scenario can be called into question based on the illogic of the Iranian regime prompting such a conflict. The chance of a successful outcome would be very small. But the Falklands/Malvinas situation remains instructive. Under grave internal pressure, the theocratic leadership might give in to its own messianic rhetoric or search for a nationalistic way of regaining popular support. Certainly a crisis in the Gulf could be manipulated to gain popular internal support. The spiraling outcome could easily be the anti-access vs. counter–anti-access conflict outlined previously.

Chapter 8

Northeast Asia: Cognitive Anti-Access and Threats of Nuclear War

In terms of regime survival, the self-styled Democratic People's Republic of Korea, or North Korea, is the most successful totalitarian state in human history. Totalitarian states seek to control the very thoughts of their population. Based on the assessments that can be made, the Kim Il-sung dynasty has exceeded its predecessors by establishing an overwhelming personality cult.[1] Even at its height, Nazi Germany was unable to eliminate such potential centers of opposing belief as the organized Christian churches. Church leaders could be co-opted or sentenced to destruction in concentration camps along with non-Aryan "subhumans" such as Jews, but the institutional framework of organized religion, a framework that taught beliefs contrary to a totalitarian worldview, was never fully subdued. With an official state religion of atheism, the Marxist-Leninists of the Soviet Union were more successful but faced the additional problem of ethic diversity—in which religious tradition plays a part. In contrast, North Korea's government appears to have destroyed religion itself.[2]

As a totalitarian state, the North Korean regime is a natural enemy of democratic governments, because the very principles of democracy argue the immorality of the totalitarian approach. In a globalized world, nearly universal access to information is the gravest threat to totalitarian governments—if other societies can succeed without such severe governmental control, what is its justification? The justification, of course, is war, which is why North Korea's regime maintains itself in a perpetual state of preparation for war—with the Republic of Korea (South Korea) specifically, but the entire world if necessary.

In order to maintain the internal cohesion necessary for perpetual war preparation of an entire society, the primary great wall that must be constructed is one against access of outside thought. This is the first sense for which it can be said that North Korea maintains a strategy of *anti-access of the mind*. A more formal term could be *cognitive anti-access*. The North Korean regime has constructed a deglobalized, counterintelligence state that can literally be more easily penetrated by a stealth bomber than it can be by a stray thought.

Whether or not such a regime can be perpetuated in the future is a serious debate, but what must be recognized is that it has survived for almost seventy years, which includes, more importantly, twenty years beyond the end of the Cold War. In the 1990s, with the democratic world heady with the thought of the "end of history" and the "inevitable" downfall of all tyrants, it was assumed that collapse of the North Korean regime was just a matter of time. From the perspective of the 2010s, that does not seem quite as clear.

Cognitive Anti-Access

Expanding the term anti-access to include cognitive factors might seem impractical, somewhat presumptive, even unstrategic. After all, anti-access was developed as a military term, a way of describing strategies or campaigns that could physically keep away an opponent who would be very hard to defeat in a battlefield force-on-force engagement. But describing mental processes or effects as a part of the anti-access concept is in keeping with the realization that successful anti-access strategies utilize diplomacy, negotiations or pseudo-negotiations, public relations, soft power, and other forms of nonkinetic deterrence or persuasion to buttress their military posture. Such nonkinetic deterrence or persuasion efforts are also the primary tools used in encouraging the *extrinsic events* that may prove distractive to strategically superior opponents. Additionally, cognitive anti-access is an obvious part of the regime's grand strategy.

For an analogy for those who might question the usefulness of the concept to strategic planning, one should think of radio silence, emission control, and stealth. Radio silence is an old term but widely recognized as referring to a posture in which all communications are avoided. Units operate autonomously, communicating (perhaps) only at close range via visual signals. The term emission control applies radio silence to all systems that transmit in some medium, such as radar, active sonar, and electronic countermeasures. The term stealth is now well known, designating systems designed not to be detected or that are minimally detected by radar or other sensors. The point of these features is to remain undetected by the enemy. It does have an obvious drawback in that the unit under such conditions cannot utilize its own active sensors but must detect the enemy by means of passive sensors or passively receive information from a distant command source. North Korea is a state that maintains tight emission control so that the outside world cannot detect what is happening internally, and so those within must rely on a command source—those around the Kim Il-sung dynasty—for their information on what lies outside their borders. This is cognitive anti-access.

The North Korean Wall

Thinking in terms of cognitive anti-access is also justified by our observation that the great walls of history have often fallen owing to internal conflict. Since totalitarian states are all about preventing even the thought of internal disagreement or conflict, they cannot allow breaches of their cognitive barriers without fearing some repercussion or recognition of the need for internal political adjustment. The extreme lengths to which the North Korean regime has gone to prevent the intrusion of the ideas from the outside world is very well documented. The implication of these efforts is that the regime perceives uncontrolled ideas to be naturally generating the internal opposition it fears. Given the nature of life within North Korea and the shockingly stark contrast with conditions in South Korea, the regime's fears would seem quite accurate.

In order to ensure that internal opposition does not develop or will not survive, cognitive anti-access is supported by the *songun* policy, literally "military first," in which national resources (as well as food) is first directed to the military before being distributed—such as it is—to civilians. Loyalty to the regime is paramount, and most important is the loyalty of the military. Although armed with aging equipment, North Korea boasts the fourth largest military in the world in number of troops, kept in readiness for a conflict with the South.

This wall is far from defensive or passive. North Korean armed forces routinely engage in provocative, often deadly incidents against the South. The torpedoing of the South Korean warship *Cheonan* in March 2010, killing forty-six sailors, is but one of the most recent. Incidents have included a commando raid attack on the South Korean presidential compound in January 1968 in an effort to assassinate the president. The Kim Il-sung dynasty ordered at least four other presidential assassination attempts, killing the first lady in 1974. The latest known attempt was in 1983 in Burma in which twenty-one people, including three South Korean cabinet ministers, were killed. In August 1976, Americans were shocked by the ax murders of two U.S. Army officers while trimming a tree in the Demilitarized Zone. The list of such incidents is a long and bloody one.

The attacks help maintain the constant warlike tensions on the Korean Peninsula, justifying the militarization of North Korean society—to which the incidents are portrayed as provocations by the South. They also make the North Korean regime feared, which is another layer in its wall. North Korea has been able to avoid retaliation by South Korea and the United States because it holds a trump card in conventional deterrence. Seoul, the capital of South Korea and its largest city, is within artillery range of the North. North Korea has the largest artillery force in the world, with about 13,000 "tubes" and 2,300 multiple-rocket launchers. Approximately 65 percent of the North Korean ground forces are located near the South Korean border, along with 80 percent of "aggregate firepower."[3] It is estimated that the North Korean military is capable of firing 500,000 artillery rounds into Seoul in one hour, which would destroy at least one-third of the city and kill thousands.[4] The warning time for the civilian population would be about forty-five seconds. Also, North Korean artillery is capable of firing chemical and biological weapons rounds. Even with an immediate military response, the South's war would start without a capital city. That fact, combined with unpredictable violence, is yet another layer in the North Korean wall.

Ballistic missiles are another offensive strength of the wall. North Korea has had a ballistic missile development program for over forty years, aggressively building its own variants from Soviet designs. It has tested short-range, intermediate-range, and intercontinental ballistic missiles (SRBMs, IRBMs, and ICBMs, respectively), and after at least one failure appears to have successfully launched a satellite using an ICBM in December 2012.[5] North Korean ballistic missiles are capable of hitting targets in most of the western Pacific region and out to Hawaii and Alaska. Satellite launches via ICBM are likely a developmental effort in increasing range to the U.S. West Coast. It is not known whether North Korea has been able to perfect a nuclear missile warhead, but that appears a clear goal.

Another layer is the protection afforded by the Peoples' Republic of China. PRC leaders may not appreciate North Korean brinkmanship, but they have consistently provided political and economic support for the regime, have attempted to dissuade retaliation, and have only occasionally permitted diplomatic condemnation at the United Nations. North Korea is not the perfect client state, but the continued existence of the regime apparently suits PRC interests.

Since 2006 North Korea has added the ultimate layer to its wall: nuclear weapon capability. Since its inventory of nuclear weapons is small and likely to remain so in the near future, this threat is buttressed by what appears irrational behavior on the world stage. The North Korean government has engaged in a pattern of diplomatic behavior that has been described as "bizarre," combining threats of war, provocative military incidents, and weapon tests with negotiations ostensibly promising a stop to further nuclear development for economic aid. Interim agreements last briefly, usually collapsing once the regime achieves but some of its objectives. This unpredictable behavior enhances the threat of its nuclear force by appearing to reject the Cold War paradigm of stable nuclear deterrence, causing potential opponents substantial concern as to how far the North Koreans are willing to escalate any confrontation. As the IISS describes it, North Korea "has effectively tried to foster the impression that it would take suicidal actions as a last resort if faced with a military threat to extinguish its regime."[6]

The North Korean anti-access wall can be summarized as consisting of:

- Cognitive anti-access: totalitarian control over its population
- Routine military provocations intended to supplement diplomatic activities
- Massive conventional and possible chemical/biological threat to Seoul
- Ballistic missile forces capable of striking regional bases and developments to increase targeting range
- Nuclear weapon development program with a small number of weapons already assembled
- Unpredictable politico-military and diplomatic behavior
- Diplomatic cover and economic support from the PRC

Weaknesses in the North Korean Wall

The primarily physical gap in the North Korean anti-access wall is air defense. Reliance on offensive systems for their anti-access posture, including nuclear weapons, is quite logical owing to weaknesses in North Korean air defenses. North Korea does possess an integrated air-defense network with hardened sites and buried communication systems, but the actual systems are older Soviet surface-to-air missile (SAM) systems with a great reliance on antiair artillery. Some of the SAM systems have had upgrades and are likely to have war reserve frequencies that have not been electronically collected by U.S. ISR systems. Yet the overall effect of the network is marginal. As an air-defense expert notes:

> The primary issue facing the [North Korean] air defense network is one of age. While the equipment may still be serviceable, none of it is a major threat to a modern air arm. [North Korea] desperately needs an infusion of modern air defense systems if it is to remain viable in the 21st century. The S-75 and S-125 [North Korea's longest-range SAMs] have been faced multiple times by modern air arms since 1990 and have consistently been defeated by current tactics and electronic warfare techniques and systems. Iraq, Afghanistan, and Yugoslavia all possessed these systems and they were all defeated.[7]

The offensive-oriented nature of North Korean forces is a likely means of offsetting the air-defense disadvantage. This posture is also "dictated by the doctrine that 'attack is the best form of defence.'"[8]

Another speculative weakness in the wall is the downward spiral of life in North Korea. Reports that the civilian population is undernourished are widespread and well documented. The North Korean regime appears almost totally indifferent to famine and other shortages of goods induced by its military-first policy. Perhaps as many as 1 million North Koreans died in the great famine of 1995–1998.[9] Countless North Korean refugees flee to China despite the fact that the PRC forcibly repatriates them to face imprisonment and torture or—depending on the number of attempts—execution. North Korea maintains the largest network of political prisons in the world today, holding hundreds of thousands either temporarily or permanently. Crimes punishable with hard labor include allowing a portrait of Great Leader Kim Il-sung to gather dust or humming a South Korean pop song.[10] As Victor Cha notes, "the only reason that we cannot claim that North Korea is the worst human rights disaster in the world today is because we are not allowed to see the extent of it." Tight government control over all information ensures that "the victims are faceless and nameless."[11]

To Western observers it seems almost incredible that the North Korean people have suffered and continue to suffer such treatment without attempting to overthrow the regime. There have been reports of assassination attempts on the Kim dynasty, even by members of their personal guards. For over twenty years, analysts of North Korea and scholars have repeatedly predicted that it is but a matter of time before the country implodes, but the regime has defied such predictions up to the present. Yet—as the history of great walls indicate—there still remains the potential that the pressure that is occurring outside the wall concerning nuclear weapon development and human rights can encourage a buildup of pressure within the wall. Moreover, the need for economic assistance and expertise from South Korea may eventually lead to an opening up of North Korean society as a by-product.[12] Whether such pressure will result in the removal of the regime, the destruction of Seoul, a conventional war in South Korea, or even a spasmodic nuclear strike is quite unknown—which is exactly the uncertainty the regime's unpredictable behavior is designed to foster. This potentially loose section of wall is plastered over with the threat of grave danger.

Cause and Course of a Conflict

Nevertheless, the most likely cause for a regional conflict is the North Korean regime's need to preserve or justify itself in the face of internal opposition. External opposition certainly has not moved the regime to change all these many decades. Even more than the Iranian theocracy, the North Korean regime views all potential internal opposition to be inspired and controlled by outside states (particularly the United States and South Korea). Adding to the situation are the facts that the Korean War of 1950–1953 has never been ended by treaty and that the North Korean regime continues to proclaim itself to be the government of all Korea as ordained by the scientific inevitability of Marxism-Leninism and the isolationist ideology of *juche*. As noted, a constant war footing is the justification for the existence of the regime, as well as its source of internal and external power.

Since it is improbable that the North Korean regime would survive a full war, particularly if it were to engage in an orgy of destruction, North Korea opts for the military provocations/incidents that are serious enough to keep up the atmosphere and pretense of war but below the threshold of eliciting a major military response. There is always the chance, however, that its calculation could be off, bringing about the major military response and the full war that it has been careful to avoid thus far. Political science literature tends to identify the precipitation of war with miscalculation, paving the way for popular interpretations about "wars no one wanted." This perception is flawed, since many of the wars described were situations in which the aggressors wanted the fruits of victory, even if they did not want the deaths of their own citizens or destruction of their own resources. The North Korean circumstances are akin to this. It is on a war footing, it has turned its society into an armed camp poised for war, it is willing to threaten attacks, and it commits warlike acts, but it sees itself immune to retaliation. If the North Korean regime pushes beyond the envelope of others' restraint, it is difficult to say it reaped a war it did not want to sow.

The two potential causes for conflict that stand out are:

- A significant threat to survival of the regime through internal opposition
- A provocative incident that creates significant death or destruction, followed by apparent outside moves to curb the regime

As feared, a war would likely begin with a North Korean preemptive attack on Seoul, South Korean and U.S. military bases in Korea, ports of debarkation, and regional bases that can support U.S. power-projection forces.[13] Ground forces would predominate in actions near the Korean Demilitarized Zone (DMZ), but ballistic missiles would be the primary tool for the first strike against bases and ports. The second tool would undoubtedly be SOF, which North Korea maintains in great numbers. Fighting initially from the defensive, South Korean, U.S., and allied forces would rely heavily on TBMD, on land and sea. The United States has chosen to develop the naval Aegis system as its primary forward-deployed TBMD system, and most of South Korea would be covered by seaborne systems—but only if the TBMD-capable ships can be swiftly deployed into position. This puts a premium on intelligence and warning of a pending attack.

The counter–anti-access phase would begin even before stabilization of combat conditions in the South, as allied strike assets (cruise missiles and tactical aviation) are targeted against WMD locations, ballistic missile sites, command-and-control centers, and North Korean troop concentrations on both sides of the DMZ. Rather than a robust anti-access defense against such strikes, allied forces would encounter the world's largest series of tunnels and buried and hardened facilities acting as a passive defense. Such facilities have been constructed throughout North Korea for almost all the years of its existence and long have been considered the major obstacle to the defeat of North Korean forces. However, some analysts now believe that a number of hardened sites, particularly ones associated with air defense, are susceptible to "bunker busting" PGMs.[14]

The priority for allied strikes would definitely be locations of WMDs and ballistic missile launch sites, since they would be the most potent anti-access weapons, although the accuracy of North Korean ballistic missiles can be questioned. It is not certain that their primary targets would end up being South Korean—or even Japanese—cities.

Another anti-access factor for North Korea would be the actions of the PRC, not necessarily for military support, but for diplomacy during and after hostilities. It does not seem in China's interest to become involved in a replay of the 1950–1953 war, and it is unlikely that it would. But it is likely that the PRC would exert diplomatic and economic pressure to "foster peace" and end hostilities prior to a complete defeat of the North Korean regime. Whether or not this effort could be successful remains speculative.

It is interesting to note that if the PRC or Russia supported South Korean/allied actions, it would enable the allies to conclude the war even quicker, since almost all of North Korean anti-access defenses and offensive forces are positioned toward the South, and attack from the Chinese or Russian axes would be unexpected.

Assessing a Northeast Asia Campaign: Fundamentals

A North Korean anti-access campaign would rely almost exclusively on offensive characteristics since it does not possess the more sophisticated anti-access weapon systems of the PRC or even Iran. The offensive nature is also a reaction to the fact that South Korea's primary ally, the United States, is so clearly *strategically superior* (1). The primary *geographic characteristic* (2) that could enable the offensive approach is the fact that, as noted, Seoul would be within artillery range of North Korean forces from the start of their operations. Other characteristics that could influence ground anti-access actions (but could also inhibit offensive actions) are the mountainous ranges that make up much of the northern portion of the peninsula and that would channel attacking forces into three major passes in any north-south movement. Since the passes could also inhibit forces moving from south to north, access to North Korea by sea would be important. Knowing this, the North Korean regime has apparently stocked up on naval mines and minisubmarines.

As in the 1950–1953 war, ground combat would be the primary mode of combat, which appears to contradict our discussion of the sea as the *predominant medium* (3). However, heavy forces would need to flow by sea from the United States, Japan, and other allied nations. Also it should be noted that it was an amphibious assault at Inchon that was the most significant strategic move of the 1950–1953 war. North Korea does not appear capable of attacking maneuvering ships at sea, giving naval forces the sort of immunity it has experienced in post–Cold War engagements. In addition to Tomahawk strikes, carrier aviation should be able to play a significant role in the counter–anti-access campaign, particularly if North Korea does successfully strike regional bases. In any event, the Northeast Asia scenario can be considered the exception to the rule in regard to fundamental #3. In fact, given the emphasis on cognitive anti-access, one could argue that the predominant combat medium would be within the minds of the North Korean population.

As in previous cases, the *criticality of intelligence and information* (4) and the use of deception would be a major characteristic. Arguably it would be the *most* important characteristic because knowledge that an attack is imminent would allow preemptive allied strikes on North Korean nuclear weapons, other WMD locations, and ballistic missile sites. This could greatly reduce the ensuing death and destruction in a modern Korean War, as well as allow U.S. power-projection forces to flow into the theater with minimum impedance. Of course, this would also require good intelligence and information on the locations of nuclear, WMD, and missile sites, and the ability to penetrate North Korean deception efforts to deny such knowledge.

One medium that might prove an advantage to North Korean forces is cyberspace, because their very limited use of the Internet and information systems in general makes them much less vulnerable to cyberattack than South Korea, the United States, Japan, and others. Reliance on space systems could also be a vulnerability for the United States, particularly if North Korea were to utilize an atmospheric nuclear blast to create EMP effects. Again, the result would have much less effect on North Korea than on the counter–anti-access forces.

Extrinsic events (5) could indeed be a factor, particularly based on the actions of the PRC and possibly Russia. Whereas North Korea seems powerless to influence events that would take attention away from its own actions, the PRC and Russia do have the potential for diplomatic, economic, and even military actions that could take global focus away from a Korean conflict, at least for a while. Conflict in the Middle East could also be a complicating factor, creating the situation of two major theater wars that was the major planning scenario for U.S. forces in the immediate post–Cold War years. This scenario has been discarded as a viable planning mechanism for more than a decade, and responding to it would stretch U.S. forces far beyond what is currently anticipated.

Victor Cha argues that—for all practical purposes—extrinsic events are what have always protected the North Korean regime from retaliation for its military provocations in the past:

> The world has watched North Korea slowly build a ballistic missile and nuclear weapons program over the past because the world can't be bothered with North Korea. The weapons are undeniably dangerous to the United States and its allies, but ultimately, an Israeli-type attack—whether in 1981 in Iraq or in 2007 in Al-Kibar, Syria—is not likely in North Korea because the issue simply does not rank highly enough in U.S. priorities. . . . Thus, when North Korea threatens, the pat response is to "park" the issue: avoid a military conflagration (because it is just not worth fighting over) by diverting attention to other important issues, and put it back on a negotiations track to prevent another crisis. This "relative crisis indifference" syndrome in Washington has saved North Korea countless times and given it benefits through negotiations rather than punishment for its misdeeds.[15]

This reliance on extrinsic events (or lack of will) as its primary international anti-access tool is disturbing not only for the outcomes of the individual provocations, but because of the potential sense of immunity it may have provided the North Korean regime. It is this sense of immunity that could prove the most significant cause of a "miscalculated" full-scale conflict.

Assessing a Northeast Asia Campaign: Supplementary Factors

Strikes against regional bases (A) and *preemption* (B) are very likely factors in the Northeast Asia scenario. Without striking regional bases, of which there are a number in Japan used by U.S. forces, there seems little that North Korean forces could do to prevent allied air operations over their territory. Strikes on regional bases would slow down the flow of U.S. forces into the region. The North Korean regime would aim for capitulation of the South Korean government or, before additional heavy U.S. forces could gain position, an armistice with its gains intact. Preemption is a natural tactic for an offensive-heavy anti-access posture that hopes to generate the "shock and awe" that might cause a wavering strategically superior power to decide that it is too costly to intervene. With a firm treaty relationship with South Korea, it is unlikely that the U.S. government would waver, but such might not be part of the North's calculus.

Preemption could also be a tactic used by the counter–anti-access force if it had certain intelligence that an attack was imminent, in order to neutralize WMD and ballistic missiles and destroy conventional forces poised to move southward. Of course, the tight control of information and deceptive measures by the North Korean regime is intended to prevent any such intelligence. Intelligence efforts versus deception and counterintelligence is an ongoing struggle on the Korean Peninsula today.

Although it extremely unlikely that North Korea could gain an edge in *technological innovation* (C) over the United States, innovation is a factor in that North Korea has demonstrated its ability to use sidewise technologies and reverse-engineering to develop its own indigenous missiles.[16] It has also benefited from the illicit network of nuclear weapon information created by Pakistani scientist A. Q. Khan.[17] At the same time, it has proliferated its indigenous missile systems, providing ballistic missiles to Iran for use as part of its own anti-access network. Iran, in turn, has tested the missiles and presumably has provided the test information to North Korea. This mode of technological innovation is what has built the most potent of its offensive weaponry.

Achieving effective *cross-domain synergy* (D) does not seem to be within the capability of North Korean forces. Undoubtedly it has created some form of cyberwarfare force, and its opponents are much more vulnerable. However, it cannot exploit space—despite recent attempts at satellite launch—and its aviation assets are no match for those of either South Korea or the United States. Its navy is optimized for coastal defense and SOF insertion, and would focus on countering any amphibious assault, but it would not be survivable in intense maritime combat. Given the relative inaccuracy of its ballistic missiles, it would be difficult to use them in direct support of ground operations. Their value is against fixed regional bases or urban areas.

In contrast, South Korean, U.S., and allied forces could achieve a degree of cross-domain synergy to allow them to coordinate defensive and offensive actions. They would need to in order to neutralize North Korea's potential for nuclear use.

Concluding Observations on the Scenario

A potential anti-access vs. counter–anti-access engagement in Northwest Asia presents stark contrasts with other scenarios. In terms of its domestic circumstances, North Korea would seem the most likely candidate for an implosive destruction of its great wall by forces within the state. On the other hand, it holds the tightest control over its population and has constructed cognitive anti-access barriers to prevent even the most benign of outside influences. In terms of longevity, it is truly the most successful totalitarian state.

Its air defenses, considered a mainstay for anti-access networks, are integrated but relatively weak, a vulnerability when facing modern combat. But it makes up for this weakness with its offensive systems—including ballistic missiles, potentially with nuclear warheads—and its ability to accept great risks in its provocative actions. With the PRC as a patron (whether a reluctant one or not), the North Korean regime appears to have the confidence that it can threaten and bluff its potential opponents while it pushes at the edges of conflict. At the same time, an obvious goal is to develop a ballistic missile capable of striking the continental United States in order to back up the threats and bluffs.

The defensive capabilities generated by the long-standing South Korean–U.S. military alliance do seem to have consistently deterred an invasion. Like Imperial Japan, North Korea could not survive a prolonged active conflict. Logically it would have everything to lose in provoking an actual war. Yet the danger of war is inherent in its more risky actions, and it is uncertain whether the Kim Il-sung dynasty truly understands the psychology of its potential opponents. Additionally, it is more than willing to portray itself as willing to engage in destructive revenge if the regime is in danger of collapse.

Whether the wall could be cracked without nuclear use by the North Korean regime is an open question and the question itself one of the most potent aspects of its anti-access network.

Chapter 9

Central Eurasia: Russia and the Near Abroad

If a military conflict between the PRC and the United States would be a war like no other, a war between Russia and the United States would be a global disaster that could end with a nuclear conflagration. Although Russia's overall military forces are greatly weaker than those of the former Soviet Union, it not only maintains a formidable nuclear arsenal but has devoted considerable resources to modernize it. The fact that overall Russian forces are much weaker is, in fact, the reason for the strategic force modernization. It is the one area of military strength that allows the Russian government to perceive itself as a great power—if no longer a superpower on par with the United States, at least equal to the other permanent U.N. Security Council members in terms of global influence.

In terms of conventional forces, however, any potential conflict between Russia and NATO (or even perhaps Russia and the PRC) would take on aspects of an anti-access vs. counter–anti-access struggle. This is not only because of the limited range of Russian power-projection capabilities, but also because the most likely sources of conflict would be crises in Russia's "near abroad" (areas of the former Soviet Union/ Warsaw Pact) in which Russian forces might seize and attempt to hold territory. Like in Abkhazia and South Ossetia—breakaway regions of the Republic of Georgia that are recognized by Russia as independent states—there are major tensions and large ethnic Russian populations within many of the now-independent former Soviet republics. Russia has proclaimed itself the protector of these Russian populations (ironically similar to Imperial Russian protection of the Slavs in the Ottoman Empire, which resulted in numerous wars).

Additionally, it is apparent that the Russian government expects the former Soviet republics to maintain a degree (or at least appearance) of subservience to Russian national interests. Thus when NATO expanded into the Baltics and Eastern Europe, Russia under the leadership of Vladimir Putin perceived it as a threat to Russia's historical influence over its neighbors—a last bastion of its conventional global influence. NATO membership for Georgia or Ukraine became an intolerable possibility that the Putin regime was determined to prevent. At the same time, Western support for Kosovo's de facto independence from Serbia, a traditional Slavic protectorate of mother Russia, was perceived as a "lesson learned": the West would support the breakaway of unhappy, ethnically repressed regions from the sovereignty of the less-democratic former members of the communist Warsaw Pact—most specifically Russia itself. Therein lies the motive for Russian intervention in Georgia's military efforts to reincorporate Abkhazia and South Ossetia. Russia could argue that the West set the precedent with Kosovo, but at the same time it could protect the two regions that had desired to remain part of a Soviet Union and had become, in effect, members of the Russian Federation no matter their status under international law. Russia has fought its own wars to retain Chechnya, a situation with some uncomfortable parallels to Kosovo.[1]

One can argue over whether the primary focus of Russian foreign policy is the eventual reintegration of the former component states of the Soviet Union into a closer association with Russia and possibly becoming part of the federation. But in any event, perpetual president or premier Putin has identified the United States—with or without the collective support of NATO—as Russia's main great power antagonist and blames it for the "color revolutions" that have threatened to replace autocratic regimes in the former Soviet republics. In the Military Doctrine of the Russian Federation (adopted by presidential edict on 5 February 2010), the first in a list of "the main external military dangers" is "the desire to endow the force potential of the North Atlantic Treaty Organization (NATO) with global functions carried out in violation of the norms of international law and to move the military infrastructure of NATO member countries closer to the borders of the Russian Federation, including by expanding the bloc."[2]

Some of these post-Soviet regimes have not been particularly supportive of Russian foreign policy or Putin's ambitions. However, a threat to neighboring authoritarian governments is inevitably an existential threat to the tightening authoritarianism occurring in Russia itself. As has been previously stated, it is difficult for an authoritarian government to justify its structure and practices to its own population when there are nearby democratic states that appear to be flourishing. As Russian journalist and Carnegie Endowment scholar Lilia Shevtsova writes, quoting retired Russian major general Vladimir Dvorkin, "what does the Russian anti-NATO faction fear? General Dvorkin had a straight answer: 'The process of joining NATO is a process of democratization that will lead to a civilizational schism between Russia and its neighbors if they join the Atlantic bloc.'"[3]

Russia's recent economic improvements are centered on resource production, which—despite its current earnings—is a very narrow base from which to create long-term growth unless great effort is made to encourage diversification. Although undertaking economic reform—primarily reduction of the political influence of some of the 1990s oligarchs—Putin's foreign policy focus has definitely been the retrieval of the great power status of the former Soviet Union. Peacetime military activities that were not conducted during the Boris Yeltsin era and are reminiscent of Cold War practices have been reestablished, such as long-range bombers' overflight of U.S. aircraft carriers during open-ocean transits, renewed emphasis on strategic nuclear ballistic missile improvements, movement of theater ballistic missiles—possibly with nuclear warheads—into Kaliningrad Oblast, and military intervention in Georgia.[4] Much of this may relate directly to justifications for greater authoritarianism and centralization of power. Shevtsova maintains that "the Russian campaign to intimidate the West, backed up with 'light artillery' on television [hostile rhetoric], has yet another goal: to lay the groundwork for a monumental distraction if the domestic situation in Russia begins to deteriorate rapidly. The militaristic rhetoric, symbolism, and pageantry (for example, the Russian navy's port call in Venezuela in 2008, or the flybys over American warships by Russian fighter jets) are clearly intended to create an enemy that Russia will bravely confront when the Kremlin finds itself unable to pull the country out of a future crisis."[5]

It is the situational parallels to the motivation of the Argentine junta's initiation of the Falklands/Malvinas War that brings us toward assessing a potential conflict in Russia's near abroad in terms of anti-access. There are, of course, important differences, the most important being that Russia is immensely more militarily powerful than 1980s Argentina and—if the conflict involves NATO members—the crisis could be on a global scale.

With this in mind, the very brief discussion that follows is not intended to "demonize" Putin's Russia or suggest that a conflict with NATO is a probability. I will let others judge that in terms of local tensions, Putin's rhetoric, and the basis for the Russian government's opposition to NATO's development of a TBMD network oriented toward Iran. Assessing a possible conflict in terms of anti-access is intended to illustrate the concept, not be predictive.

Assessing Russia as an Anti-Access State

Earlier we discussed the Soviet use of an anti-access approach as their primary naval strategy for a global war. We have also suggested that the Iron Curtain represented a political, economic, and ideological great wall. Although the current Russian regime would not describe its defense posture as being anti-access in nature, there are anti-access features. More important, Russia is the developer and supplier of many of the weapon systems that have been adopted by others into their anti-access networks, particularly integrated air defenses and cruise missiles.

Assessing Russia as an anti-access state is bound to be controversial on a number of levels. Russian government officials would undoubtedly be insulted to have the anti-access term associated with their military posture. The term implies weakness, and the Russian government has done much in recent years to reestablish the appearance of military strength on a par with that of the United States, at least in terms of strategic nuclear forces and military technological development.

Another reason is the argument that contemplating a conflict with Russia "causes" the Putin regime—like the PRC—to view the West with even more hostility and potentially take an even firmer stance against anything perceived as a U.S. objective. This view is extremely America-centric in that it assumes that public discussion in the United States somehow controls the behavior of the Russian government or that the elites do not understand Western freedoms of the press (as opposed to just not liking them). Instead the causes are internal. Shevtsova not only maintains that anti-Western actions are an effort to justify other failures of the regime, but that "there are a few more explanations for Russia's menacing growls on the world stage. The architects of Russian foreign policy are betting on two things. First, they hope they will be able to intimidate the West into giving up perceived incursions into Russia's spheres of influence. They even succeed sometimes, and if you succeed once, you're all the more likely to try again. Second, they hope that even if the West doesn't buy Russia's threats, it will still support the Kremlin's game in order to preserve good relations."[6]

Political Anti-Access

In nonmilitary terms, anti-access to the near abroad is an element of Russian grand strategy. The Russian government hopes to control the access of other nations, particularly the West, to the Eurasian states of the former Soviet Union. Economic development in the region by multinational corporations may be desired but only on Russian terms. Russia has many political and economic tools to convince the Eurasian states to limit their connections to the West. Military aid is one. Recently Russia agreed on a $1.1 billion military aid package to Kyrgyzstan.[7] At the same time, it announced a $200 million plan to upgrade Tajikistan's air defense system, probably with the objective of tying it into Russia's own air defense system.[8] Economic dependence is another. For example, almost one-half of Tajikistan's gross domestic product (GDP) comes from remittances from Tajikistani citizens working in Russia.[9]

The results have included extensions of Russian leases on military bases in the Eurasian states, and some suspect that it has also enabled it to restrict U.S. regional influence resulting from the use of bases to support logistics for the war in Afghanistan.[10] In February 2009 the government of Kyrgyzstan ordered the closure of the Manas air base to transiting U.S. forces, although it later agreed on a renegotiated lease. Settled for more money, the settlement also paralleled the announcement by

President Barack Obama's administration of its "reset policy" for better relations with Russia.[11] Washington has also been hindered in stabilizing logistical bases for Afghanistan in other Central Asian states because of its criticism of authoritarian crackdowns on human rights. Russia has seen fewer such problems in its relationship to the authoritarian-tending regimes. Returning to General Jumper's old statement that access is not an issue when one involves "the vital interests of the nation you want and need as a host," it would seem that most of the Central Asian states have concluded that their vital interests do not necessarily lie with the United States, particularly after the planned termination of Afghanistan operations in 2014.

There are two ironic developments in this. The first is that Putin himself has advised NATO not to leave Afghanistan, probably out of concern about spillover effects if the Taliban once again takes power.[12] Another explanation is that he views NATO/U.S. involvement in Afghanistan as an extrinsic event that diverts attention from elsewhere, such as Georgia or Russian human rights abuse.

The second is that in order to gain support for his efforts to keep NATO out of Central Asia, Putin has acquiesced to opening it to Chinese influence through participation in the Shanghai Cooperation Organization (SCO).[13] The PRC is likely to be even more aggressive in eventually replacing Russian influence than the United States.[14]

Right now it is Russian government policy to limit U.S. political and economic access to Central Asia, perhaps the most conflict-prone of its near abroad. The tools include military aid, economic development, access to energy, political support for authoritarian-tending regimes, and potentially including them in Russia's own air defense and anti–ballistic missile network.

Grand Strategy and Potential Scenarios

Military anti-access as but one element of grand strategy has been a theme of this book. In the case of Russia and its near abroad, the Russian government has no desire to have an actual war with NATO. It simply wants NATO and others to recognize that it considers the near abroad to be its own sphere of influence in which the desires of out-of-area powers have much less sway. It also looks for defense in depth from any imagined foe, with surrounding countries providing it—a sort of mirror image of European efforts to have a buffer to the influence of Bolshevik Russia in the 1920–1930 era. Shevtsova writes: "NATO enlargement is a threat not to Russia but to the Russian regime and elite, who want to create a *cordon sanitaire* of failed or weak states around them. For the elite, who organizes its power around the search for enemies, NATO is the optimal foe."[15]

Having failed or weak states as an anti-access buffer is a gamble, because it holds the risk of regional instability that can spill over to Russia itself. On the other hand, the benefit is that weak states are logically more pliable and can be induced to support Russian foreign policy objectives and military posture. Perhaps the greatest danger is that foreign policies of NATO or other Western states now embrace some degree of intervention in humanitarian crises in order to alleviate human suffering and restore stability. This liberal goal may prompt involvement in crises within the near abroad, particularly if the Russian government is hesitant to act or appears to be siding with forces that perpetrate atrocities such as ethnic cleansing.[16]

The desire for a buffer may confuse Western observers who remember Yeltsin-era cooperation or early Putin support for Western efforts on counterterrorism. However, closer observers maintain that such cooperation did not represent the general trend—grand strategy, if you will—of post-Yeltsin foreign policy. As Shevstova says, "the fact that Russian elites were waiting for U.S. 'deliverables' [economic compensation and end to NATO expansion] proved that they still viewed their endorsement of U.S. policy and their partnership with the United States as some kind of deviation or as a concession to the United States, but not as a strategic course for Russia."[17]

Tying together the threads of the previous discussion, we can say that the current Russian grand strategy includes:

- Maintaining the current integrity of the Russian Federation
- Maintaining internal control by current elites
- Expanding Russian influence over the current abroad
- Protecting Russian minorities within former Soviet states
- Resisting any expansion of membership by NATO
- Resisting any expansion of influence by NATO/the United States in the near abroad

- Maintaining the deterrence effect and political influence of Russian nuclear capabilities
- Improving the conventional capabilities of Russian military forces in order to have the potential for successful intervention in the near abroad
- Improving the status of post-Soviet Russia as a great power

The Russian grand strategy contains anti-access elements and logically promotes an anti-access posture concerning the territory of its perceived sphere of influence. In itself, it does not necessarily lead to an actual conflict. But a conflict scenario can arise from three possibilities.

First, a state within Russia's perceived sphere of influence may not desire to remain there. This could be a description of Georgia or Azerbaijan. If Russia were to contest their political alignments by force, this could create an anti-access vs. counter–anti-access confrontation if outside nations intervene.

Second, the Russian government might apply pressure on its neighbors in order to support an objective of Russian minority populations. This has apparently already happened in Estonia, with many experts maintaining that the Russian government conducted a "cyberwar" against the Estonian economy. This is particularly dangerous in that most of Russia's European neighbors—such as Estonia and the other former Soviet Baltic nations—are members of NATO, collectively pledged to each other's security.

Third, a potential conflict could be the result of yet another Falkland Islands–type crisis, as Shevtsova suggests. In order to divert popular attention from domestic failures, the Russian government might decide to "defend" against perceived enemies in an aggressive manner. Whatever this action might be, the relative weakness of Russian forces compared to a collective NATO would require activation of anti-access networks.

Military Anti-Access Capabilities

A complete detailing of the Russian conventional order of battle is outside the scope of this book. It is sufficient to note that Russia has a number of weapon systems that could be components of a robust anti-access network.

Foremost of course is its production of integrated air defense systems. In range and tracking capacity, the Russian S-400 air defense system is state-of-the-art. Obviously its true capabilities can only be ascertained in actual combat operations, but it appears comparable, perhaps even superior to Western systems in overall ground air defense (although not necessarily in TBMD). The Russian military is shifting from its previous S-300 systems, which remain highly desirable exports, and the Russian government has announced that the S-400 will be deployed on its southern border.[18]

As previous discussed, Russia retains ASAT capabilities and has a substantial number of satellite assets that can be used for detection, surveillance, and reconnaissance. It has outstanding cyberwarfare capability with proven effectiveness, as well as significant electronic warfare systems. The former-Soviet military was particularly practiced in concealment and deception, and there is no reason to think that such knowledge has evaporated from the Russian armed forces.

Russia also produces fourth-generation fighter aircraft and is developing a stealthy fifth-generation multirole fighter, the Sukhoi PAK FA/T-50, to rival the USAF F-22. Its military-industrial complex is but a fraction of that of the former Soviet Union, but its military engineering capabilities are competitive with those of the West, particularly in design. Its military technical innovation is superior to that of the PRC, with the result that the PRC has routinely reverse-engineered Russian designs that it has acquired.

The Russian navy is plagued by maintenance and acquisition problems, one of the reasons it has opted to purchase *Mistral*-class amphibious assault ships from France. However, Russia does maintain the elements of a sea-denial navy, with both nuclear- and diesel-powered attack submarines and a large stock of sophisticated naval mines inherited from the Soviet navy. It has recently made efforts to revitalize the navy's extensive long-range naval aviation bomber force, which was the outer rim of the Soviet Union's efforts to ensure NATO aircraft carriers could not approach its nearby waters.

Although the number of systems is nowhere near former-Soviet levels and the quality and training of its personnel remain suspect, Russia's anti-access warfare capabilities are admittedly formidable and will likely continue to increase, given current policies.

Assessing a Central Eurasia Campaign: Fundamentals

At first view, the Central Eurasia scenario does not seem to fit all the fundamentals we have used to describe the basis for anti-access strategies. It is true that in a conflict between Russia and NATO or between Russia and the United States only, NATO/the United States would be *strategically superior* (1) in terms of the quality of conventional forces. Russia does retain a formidable nuclear arsenal, probably still larger in throw-weight than that of the United States, and Russia remains a greater strategic nuclear threat than the PRC. However, we have defined strategic superiority in terms of the ability to project power from one region to another. Russia is currently unable to project sustainable military power beyond the near abroad. A power-projection military—such as that of the United States—would be strategically superior in these terms.

Geographic characteristics (2) play a role in forming the Russian military posture. Invasions of Russia from Europe have always benefited from the flat plains extending from Germany into Poland and Belarus. Russia has traditionally traded territory for time in order to allow for attrition and/or extrinsic events to sap the will as well as strength of invading forces. It is often remarked that the Soviet Union engaged in the most difficult fighting of World War II and much suffering, and so deserves significant credit for Hitler's defeat. However, it must also be admitted that without the extrinsic event of the western Allies' potential landing and return to the continent, Hitler might have been able to pour even more resources into the eastern front and break Soviet resistance. The geographic characteristics of Europe are one of the reasons that Russia seeks aligned buffer states in an effort to gain more depth.

The geographic characteristics of Central Asia are different and include mountainous regions interspersed with desert. It presents a more challenging direction of approach, with the exception of the Black Sea littoral, which provides access to the steppes—another reason for attempts to keep Ukraine out of NATO. Where geographic characteristics hinder access to the heartland, Russia can utilize smaller forces in order to hold off the enemy while it concentrates most of its forces in the more open west.

The maritime domain would *not* be the *predominant medium* (3) for much of the combat in any scenario involving landlocked states (although naval cruise missiles and aircraft could operate as in the Afghan conflict). Most of the combat would involve ground forces. However, maritime capabilities would remain the enabler in moving the power-projection forces into the region and sustaining them from a sea base if regional bases are not available.

Again, the *criticality of intelligence and information* (4), particularly through the use of satellite reconnaissance, would suggest that the conflict could spill into space as well as cyberspace. Unmanned air assets may be effective information-gathering substitutes, but Russia is probably the state most capable of shooting down drones. Of concern to both sides would be the possible destruction of the national technical means of verification (NTM) by which strategic nuclear weapon–capable systems are monitored. This could be a primary cause for a local conflict to become a global one, with frightening results. Russian emphasis on strategic nuclear weapons thereby becomes an anti-access tool: attempting access to intervene in the near abroad might trigger much worse consequences.

There are a variety of *extrinsic events* (5) that Russia could utilize to dampen enthusiasm for intervening in its near abroad, one of which is a general European dependence on Russian oil and natural gas exports. Germany and several other nations are attempting to reduce this dependence, but control over natural resources does give Russia influence over the surrounding economies. As has been suggested, Russia may encourage U.S. involvement in such places as Afghanistan in order to keep American attention away from its own actions—although U.S. forces operating closer to Russia than Afghanistan may be nearer than desired. Russia could try to fan hostilities elsewhere or bolster anti-American attitudes. It is hard to ascertain any other reason for the visits of Russian warships to Venezuela, not normally a spot of Russian military interest.

Overall, it is the *strategically superior* power-projection capability of U.S. forces, buttressed by the overall military assets of NATO, that ensures any conflict in the near abroad would take on anti-access characteristics.

Assessing a Central Eurasia Campaign: Supplementary Factors

The supplementary factors that have been used for previous evaluations are useful in mapping out the probable elements of the Central Eurasia scenario. *Strikes against regional bases* (A) are likely to be nonkinetic "strikes"—that is, political, economic, and latent military pressure to persuade regional states from allowing Western nations to have effective base access. Precrisis access to regional bases might be allowed for logistical transport to Afghanistan but without the defensive capabilities that would allow even a measure of security from attack. Such access—which is really intended to further Western involvement in an extrinsic event (Afghanistan)—could be a hostage to acquiescence to Russian activities elsewhere. Although the need for such logistical support centers may fade after 2014, the principle of providing limited access to regional bases that are, in effect, under the control of Russia is an interesting variant in achieving the objective. Like in previous examples, the front line in the anti-access effort is to ensure that out-of-the-region forces never have access to regional bases.

Preemption (B) may be less of a factor in a Central Eurasia scenario, since conflict would likely occur only after long-simmering disputes. Rather than preemption, a fait accompli that occurs while global attention is turned toward other crises could be considered the optimal tool in establishing a successful anti-access situation. If out-of-the-region powers cannot react until the anti-access state has already achieved its objective, they are faced with a calculation as to how much effort must be expended to reverse the gain and whether such effort in a marginal region—one not part of their core strategic values—is worth it. If Imperial Japan had simply moved into former French and Dutch possessions and avoided U.S. or British territories, would the United States have declared war on Japan—even if it knew that it would eventually win any prolonged conflict? That is a question to which historians might turn their attention.

From the perspective of a detached, clinical observer, the *technological innovations* (C) applied in the scenario would be rather interesting. NATO members and Russia are the world's top innovators in military technology. It has long been surmised that NATO maintained a qualitative edge over Soviet-era military technology, which Soviet forces mitigated through quantity of systems. Although the West does appear to possess the edge today on both accounts, it is difficult to ascertain in peacetime how wide or narrow the edges might be. U.S. naval strength—based in part on advanced technologies—remains superior to that of Russia, but that would seem natural given the geographic characteristics of both states. U.S. ground force technologies appear superior, supported by clear advantages in training and personnel qualifications. Russia remains much more competitive in missile and aircraft technologies, both traditional strengths from the Soviet era.

As in the East Asia scenario, a full-scale anti-access conflict would take the character of attrition warfare using high-technology weapons. Under such circumstances, the side that most effectively blends tactics and technologies would have the advantage.

Such a blend is dependent on *cross-domain synergy* (D) because full-scale combat would cross all mediums and domains. Preventing destruction of civilian infrastructure while maximizing military striking power might be a particular feature of this scenario, since it would be logical for both sides to want to keep actual combat localized in the near abroad and away from the populations and core infrastructures of the homelands. To achieve such restraint requires particular skill in coordinating cross-domain operations, ironically applying limits in ways that do not completely dissipate the overall synergy.

Concluding Observations on the Scenario

Including this scenario in a study of anti-access might seem quite a distance away from the particular assumptions of the AirSea Battle concept or other sustained discussions of anti-access challenges. But any conflicts in the Russian near abroad that involve outside powers or alliance networks such as NATO will inevitably have anti-access characteristics. The Russian government would seek to avoid the potential for outside interference and limit access to the region in any way that it can. Prior to hostilities, this would involve diplomatic, economic, and military assistance activities, once more highlighting the fact that countering anti-access is not simply a military problem.

This scenario does not follow the analytical patterns for anti-access strategies that we have established through the evaluation of other examples, primarily in *the predominance of the maritime environment* (3) as combat space. But that is dependent on the size and scope of the conflict. Any near abroad crises that escalated into a major NATO-Russia conflict *would* involve the maritime environment as a combat space, as well as a primary medium for access to the Russian periphery. Perhaps the inclusion of the scenario will prompt other services to look at the anti-access problem in new ways and not simply focus on the narrow, traditional confines of area denial.

Whether the scenario escalates into a much broader conflict is largely dependent on which countries Russia considers appropriately in its sphere. The conflict with Georgia did not broaden because Georgia was not a NATO member and NATO/U.S. attention and commitment remained elsewhere. Most of the Central Asian states are not well connected to the overall global economy and have not developed formal alliances outside their regions. But a crisis that involves the Baltic republics—all NATO members—whose economic successes seem a challenge to the Russian model and who contain often glum Russian-speaking minorities, would broaden swiftly. If critics of Russian policy such as Shevtsova are correct that the primary motivation of the Russian government in portraying NATO as an enemy is to ensure that there will be a distraction in time of domestic failure, a Falklands/Malvinas-type crisis is easily envisioned. Such a possibility deserves serious analysis and not simply for its logical anti-access characteristics.

Chapter 10

Breaking Great Walls: Issues of Modern Counter–Anti-Access Strategies

It is more satisfying for both author and reader to conclude with recommendations rather than questions. But there is no shortage of specific recommendations concerning what the U.S. and its allies and partners should do about anti-access warfare. Rather than simply add onto the pile, this book—taking more of a scholarly than advocacy approach (for better or worse)—concludes with some issues for consideration. Discussion of these issues, however, should provoke recommendations or at least some thinking about options for policy. The questions that are asked are posed in the context of both the conceptual arguments and past examples of anti-access and counter–anti-access strategies.

Lessons from History

The emphasis in this book has been on understanding the concept of anti-access in respect to the lessons of history in order to try to apply these lessons to modern scenarios. This is by design because it is an approach that has not been done before. Studies of modern anti-access scenarios have cited historical examples, but they have been poorly developed and in some cases are simply poor examples meant to fill space. Cynically one might say that everybody likes to cite history, but few really want to study it in detail. This book does *not* study the cases in the detail they deserve, but hopefully it will inspire others to look at these and similar campaigns, and determine whether the anti-access concept truly describes the strategies involved.

It is recognized that some analysts of modern anti-access scenarios will dismiss this approach because—well—it's old history. Who really cares that the ancient Greeks used an anti-access approach to defeat Xerxes? What does defeating the Armada with short-range and highly inaccurate cannon have to do with antiship ballistic missiles that can range a thousand nautical miles? Doesn't technology trump history?

Modern Anti-Access Strategies Represent Evolution, Not Revolution

In historical terms, changes in military technologies are rarely a revolution in military affairs. They are evolutions in military affairs. They are interpreted as revolutions from hindsight, similar to the way historians of the 1800s determined that there had been a Renaissance in Europe. Leonardo da Vinci, for example, did not know he was in a Renaissance; he simply tried to innovate designs and improve existing things.[1] Those who write about past revolutions in military affairs often make it seem that the changes occurred rapidly and that those involved recognized the obsolescence of past practices and discarded them (or foolishly clung to the obsolete practices). Those who point to such examples as tanks making cavalry obsolete give us the impression of obstinacy, fixation with tradition, or downright stupidity on the part of those wanting to retain the technologies of the past. There has always been some of all three in military affairs. However, in many cases older technologies did not instantly fade away, and for good reason—they proved useful under circumstances for which the new technologies are not optimal. They are also useful when a certain number of systems inevitably fail during a full-scale battle in which the enemy tries to destroy those systems. Those who push for rapid change and are willing to rapidly discard "legacy" technologies are not always giving us good advice.

Returning to the case of cavalry, an often cited case of obstinacy in the face of new developments is that of Polish cavalry charging German tanks in World War II. It never happened. The idea started as Nazi propaganda. What really happened in the infamous incident was that Polish cavalry drove off German infantry, and then the infantry unit called in the panzers to drive off the Poles. The Poles wisely retreated into terrain in which the tanks could not follow. Not to flog a horse, the Germans themselves—although fathers of the blitzkrieg—used cavalry very extensively on the eastern front and in partly mountainous areas where tanks could not operate. In fact, the Germans intended to bring horse-drawn field artillery across the Channel in Operation Sea Lion. Draft horses weigh much less than tanks and trucks, and are easier to load onto small barges. It was old technology that could do something that the new technology could not do. Unfortunately for Poles, much of their territory is flat plain on which tanks could operate. In mountainous Afghanistan, U.S. Special Forces personnel rode horses with the Northern Alliance in initial campaigns. New technologies have to fit the circumstance. Rather than revolutions in military affairs, there are evolutions in military affairs, and this applies to anti-access strategies.

Technology Not the Sole Answer

Why is spending time with this evolution-versus-revolution debate important for understanding anti-access strategies? There are at least two reasons. First is the fact that for every technological advance in anti-access systems, there is invariably a countermeasure. Antiship ballistic missiles could potentially strike, possibly even sink, an aircraft carrier, but the joint force could put deception systems on every ship to imitate an entire fleet of carriers. Does the enemy have enough ASBMs to target every vessel, all of which are moving? Much has been written—mostly in media reporting and on op-ed pages—that implies that aircraft carriers are obsolete in the face of anti-access systems.[2] They might become so if we are not smart. To be smart, we have to study anti-access networks seriously from a tactical perspective and not be fixated on the threat (or acquisition) of new technologies.

There *are* new technologies. They must be studied. Countermeasures to them must be developed. There *are* significant issues, such as the fact that the operating ranges of carrier-based aircraft have actually shrunk since the Cold War and having to operate the carriers closer to land can make them more vulnerable. But to continue the carrier example (because it applies elsewhere), that does not make the concept of carrier operations obsolete; it calls for naval aircraft (and UAVs) with greater range. In the face of modern anti-access networks, all tactical aircraft need greater range. Discarding so-called legacy systems or concepts without first trying to operate them in new ways and with new improvements is not logical. Greater range might be achieved with a different mix of aerial refueling capabilities. "Transformation" can be a good thing. It can also be bad. In comic books and movies, Bruce Banner transformed into the Hulk and became much stronger, but this was not a good thing. The sky is not falling because of modern anti-access technologies.

The second and even more important reason is that, as the historical examples indicate, military technology is rarely a decisive factor in the anti-access/counter–anti-access struggle. Radar did provide RAF Fighter Command with an advantage over the Germans in the Battle of Britain. The Germans misunderstood the importance of radar to the overall anti-access network and suspended the strikes that were specifically conducted against the radar sites. Airfields were always a more important target to them. Radar itself did not win the battle or eliminate the threat of Operation Sea Lion. It was German reluctance to press the attack—a reluctance that they hardly showed elsewhere—in the light of extrinsic events (the desire to focus on the eastern front) that eventually determined the result. The Luftwaffe suffered badly, but so did the RAF. Much worse for the Germans was their continuing faith in their cryptologic technology, a development that failed them. The Allies' ability to read Enigma dwarfed the importance of aircraft characteristics. The effective dissemination of SIGINT to the operational commanders and the Double-Cross System of turned German spies represented brilliant operational successes rather than advances in technology. With the exception of atomic weapons, it was the Germans who made the greatest advancements in military technology—witness their V-1 and V-2 missiles and jet aircraft. Fortunately they appeared too late to significantly strengthen the German anti-access network. The wall was already breaking.

Commitment and Extrinsic Events

What *has* beaten counter–anti-access efforts is not weapons or technological advancements or innovative tactics. Rather, it has been a wavering of the out-of-area state's commitment to the operation owing to a concern for extrinsic events.

In the case of Gallipoli and the Dardanelles, the British-led forces were in a difficult situation. Part of this was owing to a lack of initiative on the part of the field commanders. But it was not simply the casualties that caused the war council to fold. Many more soldiers were sent to their deaths charging against entrenched machine guns on the western front. The Turks could not push the ANZAC and other troops into the sea. The ships sunk were not first-rate. Younger, more energetic naval officers had innovative plans to renew the naval assault in the Straits. British submarines were operating in the Sea of Marmara.

What caused the British to make the decision to desist was the fact that the counter–anti-access campaign against Turkey was a sideshow and not in what they viewed as the decisive theater. It was a good idea if it could be done swiftly. From this perspective, the casualties and lack of progress were not worth it. Given the routine political instability in Ottoman Turkey, the Turks were being pushed to the breaking point. They had active plans to burn down Constantinople rather than let it be captured—hardly a sign of confidence. The British did have fears that German supplies (and possibly more personnel) would be shipped to reinforce the Turkish forces, and some artillery did arrive there when the campaign was over. Yet a creative mind could have argued that German supplies to Turkey meant fewer flowing to the western front. Perhaps Churchill did argue such. Nevertheless, the potential results of a successful, but prolonged campaign were just not that important to the British decision-makers. In contrast, the Allies broke the walls of Fortress Europe and the Japanese Empire with a will that could not be dissuaded by the length or costs of the fight.

We have argued that counter–anti-access campaigns often turn into battles of attrition as well as maneuver. That is certainly an appropriate depiction of the Allies' counter–anti-access efforts in World War II. In the modern setting, this might not mean attrition in large numbers of personnel, rather attrition in weapon systems, sensors, and platforms. Million-dollar missiles will be fired off in the hundreds. Aircraft and ships will be lost. Satellites will be destroyed. One must ask whether the commitment to take apart modern anti-access networks—to bring down the wall piece by piece—will be present in "wars of choice." That is a question this book cannot answer, but we all have our suspicions.

Counter–Anti-Access and Deterrence

Suspicions that a commitment to counter–anti-access campaigns is questionable in the modern scenarios do not necessarily rule out the necessity for investing in counter–anti-access systems. In 2012 President Obama made a commitment on paper to do so. This makes sense from the perspective that the primary purpose of the U.S. Armed Forces is not to fight wars, but to deter war.

The requirement for nuclear deterrence is still a factor in international security, particularly with potentially messianic states such as Iran and North Korea in possession of nuclear weapons. We don't really know if deterrence or forbearance has prevented nuclear war. We do know that there hasn't been one. Being logical, we must assume nuclear deterrence worked in the Cold War and probably does today.

On the issue of the effectiveness of conventional deterrence, there is doubt. Conventional deterrence—that is, deterrence maintained by nonnuclear weapons—is fiercely debated. Conventional wars have occurred even when one of the potential opponents appeared to have a preponderance of destructive power. In the case of Saddam Hussein's invasion of Kuwait and the Argentine junta's invasion of the Falklands, it is apparent that neither expected a military response to their actions even though their potential opponents possessed superior forces.

Yet based on historical experiences concerning the outbreak or prevention of war, we must assume that conventional deterrence has some degree of effectiveness.[3] Even without resorting to game theory or elaborate modeling, the property of deterrence is eminently logical and has everyday domestic counterparts. It is well known that effective law enforcement has a deterrent effect on crime. The often quoted passage from the Roman strategist Vegetius that "those who want peace must prepare for war" has been handed down through the ages because it is what many, perhaps most people believe. And belief is the mental basis for deterrence.

In practical applications, one asks about the deterrent effect of counter–anti-access capabilities under modern conditions. If, for example, the PRC thinks that U.S. forces could penetrate anti-access networks and prevent it from a clean victory in forcibly annexing Taiwan, would it choose to risk an attempt? If the PRC thinks that U.S. forces could *not* penetrate anti-access networks and would *not* be able to operate effectively past the two island chains, would it choose to try? What if the CCP is under tremendous pressure from a sinking economy or domestic unrest and desires to demonstrate some sort of legitimacy? Intention is a difficult thing to predict, but it should occur to even the most casual observer that the second belief (U.S. inability to penetrate)—even if it is inaccurate—provides greater incentive to act than does the first (U.S. ability), if forcible annexation of Taiwan is a firm goal.

On the other hand, deterrence involves much more than military strength. What effect does the economic relationship between the PRC as manufacturer and the United States as consumer have on the CCP's calculus? Would the PRC's forcible annexation of Taiwan result in the loss of access to the global market? Is Taiwan worth that? Is intervention against such an action in the economic interests of the United States? More importantly, do CCP decision-makers think that the U.S. government does or does not view intervention as being in accord with its economic interests? Nevertheless, in the face of such questions, it remains logical to assume that the ability to counter and penetrate an opponent's anti-access network has a deterrent effect on any decision to go to war.

In the cases of Iran, a more authoritarian Russia, and a xenophobic North Korea, deterrence may require different tools. Economic sanctions appear to have effects on Iran but not on North Korea. Skilled diplomacy may be the most valuable tool in negating Russian tendencies toward regional anti-access policies, because it appears that the Russian government's foremost objective is to maintain its appearance as a global power, no matter the behind-the-scenes regional activities. A conflict centered on Iran has different implications than one on North Korea. And a conflict with the PRC or Russia may have dire global results.

As we have noted, viewing deterrence as the primary purpose for developing and maintaining counter–anti-access capabilities is problematic for efficiency experts and budget "bean counters." Investments need to be made in the hope that the resulting acquisitions are never used in combat. That is true of all military acquisitions and particularly true during periods of economic downturn and for systems optimized for "attrition with precision weapons" scenarios. Deterrence must be credible, but since no one really knows what makes it credible, debate will always ensue. A prudent strategist might argue that the most credible deterrent is the latent capability to defeat the enemy—in this case, break through the anti-access network. But that is often a hard sell, particularly concerning conflicts that "no one wants."

Anti-Access as an Element of Grand Strategy

Describing anti-access as an element of grand strategy may provoke a degree of push-back on two accounts. First would be the reaction of those who view countering anti-access networks as a practical military problem best solved by analyzing tactics and orders of battle. Part of this crowd would acknowledge that it *could* be a part of an opponent's grand strategy, but the discussion takes away the focus on what is truly important—such as military technological developments. On the more extreme edge, some would view discussions about grand strategy, the interrelationship of economic security (access to raw materials and markets) and defense posture, control of global commons, and so forth as so much "globaloney" that has little to do about hard-and-fast intelligence on whether an ASBM can sink a carrier or whether we really need to spend money developing a V/STOL variant of the F-35 (the F-35B).

Leaving aside the debate on whether intelligence is ever "hard and fast," one can reply that concepts that do not have a strong analytical basis, including historical evidence of validity, tend to get pushed out of the way by the even more recent defense buzzwords (witness "effects-based operations"). Describing and analyzing anti-access strategies as elements of grand strategies provide the basis for understanding why a state would invest resources in an anti-access network in the first place. What are its core objectives? Why does it need a wall? Without looking at the concept from a broad perspective, it is easy to be trapped in the long-term situation ascribed to many American military campaigns: winning the war but losing the peace.

Authoritarian regimes naturally tend toward building both military walls and walls around civil society because survival of the regime itself is considered the primary security objective. Both walls are needed if there is the perception of threats both within and without. Walls provide a sense of security from outside interference, whether military intervention or the spread of discomforting ideas.

A second group that might push back consists of those who believe that true coordination between the elements of national power is unachievable in a democracy (and perhaps in the Byzantine jealousies of many autocracies). How can one have a grand strategy if the Defense Department wants to restrict the export of technologies that might be used in anti-access networks and the Commerce Department is pressing to support the companies that want to export it? There are those in the State Department who truly believe that diplomacy is always an alternative to war, rather than an activity that might contribute to either war or peace depending on circumstances. Then there is always the inertia on the part of political leaders—and sometimes bureaucrats—to delay making a decision in the hope that some other development (and there are always new developments) might render an immediate decision moot. American (or multinational) companies dependent on access to the low-wage Chinese work force or whose business consists of selling Chinese exports in the U.S. retail market do not want their (or any) government "demonizing" China," thank you. In this cacophony, why bother worrying about the problem in terms of grand strategy? After all, isn't America's traditional grand strategy one of muddling through?

This book emphasizes anti-access warfare as a part of grand strategy in the belief that without applying the other elements of power—political, economic, diplomatic, and so-called soft power—you cannot effectively defeat robust *military* anti-access networks except at great cost. Access to regional bases is an obvious requirement that necessitates effective diplomacy, a feeling of shared political objectives, and perhaps economic incentives, but there are many other nonmilitary "phase 0" requirements. The Defense Department has only the kinetic part of the problem.

How many diplomats recognize the concept of anti-access? Who outside of DOD (and for that matter, how many purported strategists in DOD) have actually read the *Joint Operational Access Concept*? Who has the time when focused on the immediate problems of a specific department? Contemplating the answers to these questions naturally drives one into the "grand strategy is muddling through" camp. But that again confines anti-access warfare to being a military-only problem, which is where it presently resides.

Anti-Access Is Not Just a Military Problem

If anti-access *is* an element of grand strategy and grand strategy is defined as entailing all actions intended to preserve a state's freedom of action in order to achieve its objectives, we can see that anti-access strategies are not just a military problem. "Whole-of-government approach" has become a cynical cliché. But—as stated—the deterrence phase of a counter–anti-access campaign requires the use of diplomatic and economic activities that are not under the control of the Department of Defense. The support of regional allies must be maintained in order to ensure the use of air bases, land force bases, and port facilities, access to prepositioned military equipment, and possibly combined operations with the allies' armed forces. The political groundwork for such support requires diplomatic activities culminating in treaties, access agreements, and status-of-forces agreements. Following establishment of the legal basis of the relationship, periodic and routine diplomatic effort must be made to maintain the relationship through reassurance efforts such as high-level discussions and public diplomacy. Support must be demonstrated for related and unrelated political objectives of the allies, such as supporting votes at the United Nations or in the proceedings of other international forums. Much of this support must be for objectives in which the U.S. government may have no direct interests or possibly mildly contradictory ones.

Stable, mutually beneficial economic relationships must also be maintained with regional allies. Economic relations usually have the most direct effect on the peacetime well-being of the citizens of the allies and partners. Most-favored-nation status, free trade agreements, and other economic agreements or even concessions are typical methods of creating and maintaining a favorable economic relationship. Not only does this require cooperation in the U.S. government by the Treasury Department and the Commerce Department (and other entities) with State and Defense, it also requires a spirit of cooperation, support, or acquiescence within the private sector business community in arrangements that might cut into its global competitive advantage. There are strong public diplomacy and strategic communications components in this. As implied previously, there also may be constant tension within the U.S. government concerning the balance between domestic business interests and the economic activities that may be needed to keep the politico-military relationships in the region.

An example of positive intragovernment (and international) cooperation is the successful counterterrorism effort. This example works us away from the muddling-through camp.

Anti-Access Requires a Joint Solution—Not Jointness for Its Own Sake

Returning to the military component, there have been comments throughout this book concerning a growing ideology of jointness that appears to prize equality over effectiveness. This has been true in recent years concerning novel concepts (and buzzwords) such as "effects-based operations," "irregular warfare," and "stability operations." And this has become particularly evident in interservice discussions of anti-access and AirSea Battle. The creation of the joint USN-USAF AirSea Battle Office has earned much praise from the Office of the Secretary of Defense but has generated some concerns on the part of the other services. Conferences have been held on how they can get their "share" of AirSea Battle, and many commentators have suggested that all services need to have equal representation on every such issue.[4] In particular, U.S. Army planners have felt left out because AirSea Battle seems—logically—to have no call for large formations of heavy land forces or for counterinsurgency forces. As discussed, implying that a particular service would have a limited role in a major war-fighting concept is not viewed as particularly joint—which is why the subset (or perhaps redundant) term area denial is regularly tacked onto anti-access. It is hard to distinguish area denial from routine *tactical* objectives of land warfare. But tacking it on certainly gives the impression of service equality. No one is left on the bench.

The problem with the current ideology of jointness is that it drives planning to the lowest common denominator of strategy. Seamless joint operations are essential for combat, and the overall jointness movement that resulted in the Goldwater-Nichols Act of 1986 has led to the prioritization of service interoperability and has made a tremendous improvement in U.S. military effectiveness. It was needed. Yet it must be recognized that although all services may participate to some degree in counter–anti-access operations, the levels of contribution will not be equal. The resources required will not be equal. Thus the allocation of defense resources, as it affects preparations to conduct a counter–anti-access campaign, *should* not be equal. Levels of contribution will change based on the scenario.

Countering anti-access warfare is not about just fighting a naval and air war. It is about defeating an opponent whose strategy is to neutralize a superior force until time, attrition, and/or extrinsic events shake the superior force's determination. The force must be tailored to the task based on the scenario. That means in some scenarios part of the joint force will *not* be used. In the East Asia scenario, it is hard to see how land forces can play a role unless the intent is to garrison Taiwan. In the event of Taiwan being conquered by the PLA, a recovery might be attempted, in which case robust amphibious assault/forcible-entry capabilities would be needed. Of course, if the conflict goes to that point, there would be many more effects and challenges. But the potential for a clash of heavy land forces in mainland China is currently not plausible.

In the Southwest Asia scenario, land forces and SOF would likely be required for capturing littoral areas that could threaten maritime operations and freedom of transit through the Strait of Hormuz. That is the traditional role of the U.S. Marine Corps, which has always been a light expeditionary force. Again, it is hard to envision a role for heavy land forces unless the goal is regime change and the garrisoning of Iran—which would make the Iraq intervention look easy in comparison.

In the Northeast Asia scenario, heavy land forces *would* be heavily engaged. But the anti-access discussed in that scenario is not about a conventional invasion of the South to gain territory, which has been effectively deterred through the stationing of U.S. forces alongside those of the Republic of Korea. Access to the North is needed to prevent the most frightening threat of a *Götterdämmerung* attack if the collapse of the regime is a prolonged, bloody affair. That is not what we consider a conventional scenario for the fighting of a land war.

The Central Eurasia scenario is another that would appear to primarily feature land combat. But how would ground forces get there? Where would they be based? The former Soviet Union had a very sophisticated sea-denial strategy to prevent NATO forces from penetrating its periphery. If the Russian government were convinced that U.S. forces would engage them directly on the ground, would it allow U.S. forces to enter the region? On the other hand, if it were convinced (or even suspected) that U.S. forces could penetrate its anti-access network, would it be enthusiastic about flexing its muscles on neighbors that might appeal to outside aid? In that case, wouldn't there be a disincentive to apply military force?

The point is that in all the potential anti-access scenarios, equal application of the joint force is *not* what is required. In fact, much of the joint force cannot be applied if decisive operations have not already occurred in and from the maritime areas that lead to the region. The maritime domain (again, including the air and space above it) is the entryway in the anti-access vs. counter–anti-access campaign. This is a fact of global geographic characteristics, not some desire of addled navalists. To fight in the region, one has to enter the region. The enemy does not want to fight *in* the region. If joint forces are to be tailored for the counter–anti-access struggle (through acquisition programs, organization, tactics, and so forth), they must be tailored with that in mind. It must be acknowledged that, as concerns the United States defeating anti-access warfare, successful maritime operations are a prerequisite for joint operations—not an add-on, not yet another domain, not just one of a number of equal claims on resources.

Information and Intelligence Are Critical, but the Fog of War Is Unavoidable

Common to all scenarios is the need for information and intelligence in order to understand postures and possibilities. Anti-access forces that are deceived by the enemy as to its actions or intentions must have a depth of capabilities that allow them to recover. Nazi Germany did not, with its forces tied up in the east and its eyes in the west focused on the wrong beachhead—its preferred solution. This is also true of counter–anti-access forces. In the effort to force the Turkish Straits, Britain deceived itself as to Ottoman resilience. In contrast, Elizabethan England had good knowledge of Spanish intentions and capabilities, and could deploy its forces effectively—even achieving a preemptive attack that delayed Spanish naval movements into the region. Yet uncertainty could never be dispelled, which meant redundancy in force capabilities and the ability to adapt were always premiums.

Today's world is awash in information because there are new technologies that can rapidly store, correlate, and disseminate it. Yet a good deal of that stored, correlated, and disseminated information is not necessarily true or is only approximately so. It was fashionable in the 1990s to suggest that emerging information systems promised to lift the fog of war, although that view has suffered in the wake of recent conflicts. In anti-access warfare, the goal of both the anti-access and counter–anti-access force is to create as much possible fog.

Counter–anti-access forces must operate under the assumption that they will face much fog. At the same time, they must try to prevent it through self-protection of sensors and information infrastructures. If ISR sensors can be protected, for example against ASATs and EMP, and computer attacks defeated, then deterrence is enhanced. To do this, hardened sensors and computer network defense must be a priority.

The cyber dimension is a vital operational domain for high-technology forces. Cyberattacks and cyber defense have not been discussed in any great detail in this book since there have been extensive discussions elsewhere. A logical conclusion from these discussions is that anything connected to an outside network is vulnerable. Even prioritizing defenses does not mean we can assume they will work in a counter–anti-access campaign. Unfortunately, this realization bumps up against cost and convenience. It is convenient to assume that information systems can be made secure, just as it is convenient to assume regional base access will always be available. Developers now market cryptologic software they claim is unbreakable. The Germans thought Enigma was unbreakable too.

Since there will always be vulnerabilities as information warfare evolves—hacking versus patching—an option could be the creation of a small, relatively low-tech/low-observable counter–anti-access force that can initially penetrate the outer rim of anti-access networks as the rest of the power-projection forces reboot. To create such a squadron/wing/group/division would require means of operating in the absence of GPS, broadband (or possibly any) communication, or other long-range connectivity. This is not the direction that the joint forces have been traveling, even if special operators have learned to ride horses.

A known ability to operate effectively in and through the fog of war could have a deterrent effect on anti-access states. Deterrence against aggressive actions by an anti-access state is logically most effective if the anti-access state can never assume that its wall is impermeable, even if it can get inside its opponent's information loop.

Sensors over Shooters as the First Targets

The reverse side of acquiring information, protecting it, and being able to fight through the fog of war is destroying the enemy's sources of information. It has been collective wisdom that in high-technology warfare, one should concentrate on striking the archer rather shooting down the arrows. Without an archer, arrows cannot be shot. For counter–anti-access warfare, it should be emphasized that the first targets are the eyes of the archer.

This is the reason why space and cyberspace are significant mediums for anti-access conflict. ASAT weapons have been emphasized in this book because they are the most direct means of destroying an enemy's space sensors, particularly those that have been hardened against EMP. Other techniques, such as lasers, could also be used to blind or dazzle space sensors. Research and development of a variety of such systems is undoubtedly under way. At the same time, use of ASATs threatens the NTM, considered critical elements of the strategic nuclear deterrence equation. Deciding to strike space sensors would be a difficult decision against an anti-access foe with an extensive strategic nuclear arsenal—Russia in particular and perhaps the PRC soon.

Attacks on ground- and air-based sensors are also imperative, perhaps with less worry of nuclear escalation. In any event, striking the sensors of any anti-access network—be they in space or on earth—is logically the initial stroke of a counter–anti-access campaign once active hostilities begin. Attacking the shooters comes after destroying the sensors.

War of Attrition: Quantity and Redundancy Are Key Requirements

Self-protection and strikes on sensors would not be conducted solely in the initial stages but—like attacks on the shooters—throughout the conflict. It is important to emphasize that an anti-access vs. counter–anti-access struggle is likely to turn into a war of attrition as well as maneuver.

The words "attrition warfare" evoke horrendous images: trench warfare, Gen. Ulysses S. Grant at Vicksburg or Cold Harbor, or relative indifference toward the lives of troops or civilians. It has become a forbidden word in the defense dialogue. All joint planning assumes that maneuver is always superior to attrition. In fact, references to attrition cannot be found in most joint documents.

Yet attrition is part of war and a natural element of an anti-access struggle. It cannot be ignored. In recent wars, opponents have aimed at maximizing U.S. personnel casualties not to achieve particular tactical objectives, but to break American will to continue. More than anything, they wanted the Americans to leave. This is a method suitable to anti-access warfare where the objective is to neutralize the superior force until time, attrition, and/or extrinsic events shake its determination. In "local wars conducted under high-technology conditions," attrition is not specifically aimed at personnel casualties, but at the destruction of weapon systems. Once the opponent is out of missiles and attack aircraft, it cannot engage at long range. If regional bases are destroyed, there is nowhere for shorter-range forces to marshal. If all or most ships are sunk, it is impossible to move heavy forces into the region. The anti-access state wants the war to terminate at that point. Its objective is to eject the superior power from the region, not conquer it. The calculation is that the expense and time of building new high-technology systems to continue the fight will convince the strategically superior counter–anti-access power to quit because the regional objective is not worth the cost. It is the Imperial Japanese gamble, which is appropriate because the Pacific War is the most intense model of an anti-access vs. counter–anti-access campaign. It also explains why preemption is such an attractive technique.

Viewed from this perspective, preparation for a counter–anti-access campaign cannot consist of building a small, very expensive inventory of precision weapons. The more precise a weapon, the more operationally effective it is. But it is the robustness, sustainability, and survivability of the overall force—with built-in redundancies, which can act as alternatives when specific capabilities are neutralized—that are the requirements for effective counter–anti-access. This is an unspoken and largely unrecognized assumption when the term *cross-domain synergy* is used. Precision weapons will be needed in building counter–anti-access force quality but must be balanced with quantity in order to achieve effective use of acquisition resources. Weapon systems must be able to conduct sustained attacks, not just precise attacks, in order to defeat a high-technology anti-access network.

This is an awareness that has not yet become a part of the AirSea Battle dialogue. And it is certainly not a part of the DOD's acquisition trends. If we truly believe that anti-access warfare characterizes the wars we will face in the future, acquisition trends will need to be changed. Identifying a single point of light, an effective counter–anti-access force needs arsenal ships that epitomize Hughes' maxim of "attack effectively first"—not lightly armed, nonsurvivable LCSs. Also, the systems needed do not—cannot—be "gold plated." Instead, they need to be rugged and have built-in redundancy.

Preemption Is Always a Factor

As discussed previously, preemption is alluring. Anti-access states will always be tempted to preempt if they have the means to do so. Imperial Japan is of course the best example. The anti-access state does not want to allow the strategically superior power to have the time to assemble its forces and proceed in a methodical fashion. It certainly does not want residual opposing forces within the region. Protracted conflict can be a factor in breaking the strategically superior power's will—but only if the anti-access state is *winning*. Allowing all the chess pieces on the board is not a likely avenue to victory.

Some might dismiss preemption as unlikely and illogical. The results could be escalation to a conflict focused on revenge and regime change. But cold logic does not always apply to war or its planning. And sometimes cold logic is strategically counterintuitive. As noted, Admiral Yamamoto studied at Harvard, had two tours as a naval attaché in

Washington, traveled in the United States, and knew that Japan could not defeat the United States in a long war. He had faced assassination attempts by fanatical officers who considered him pro-American. Yet he still planned the Pearl Harbor attack. As a professional, he knew that it was the necessary foundation for the anti-access strategy that was Japan's only hope for victory against the United States. It was the only move that could conceivably prevent the strategically superior power from reentering the region.

In the contemporary scenarios, preemptive strikes on regional bases—and perhaps even out-of-the-region ports of embarkation—represent cold operational logic. Preemption cannot be dismissed. It must be planned for if it is to be deterred.

Sea Basing of the Counter–Anti-Access Force

One possible solution to the vulnerability of regional land bases is sea basing. Sea basing is a concept that has been pushed by the U.S. Marine Corps, with modest U.S. Navy interest. In essence, sea basing refers to the capability of using the sea in the same way that U.S. forces use overseas regional land bases for deterrence, alliance support, cooperative security, and other forward operations.[5] Sea basing is not necessarily about building large platforms at sea, although some have interpreted it that way. In the 1990s a Joint Mobile Operations Base—consisting of offshore oil well–type platforms attached together—was proposed. However, the sea base could also consist of dispersed platforms netted together by a common operational picture (common access to tactical information), which operating forces could combine as needed.

The first advantage of a sea base for the counter–anti-access force is that it allows the force to be stationed in the region without the issues of sovereignty and permissions required of base operations in other countries. The second advantage—perhaps more important—is that it can be constituted to move, making targeting more difficult for the enemy than a fixed land base. Unfortunately, the concept energizes a lot of debate over service roles and missions, and has been a potential joint program with little joint support. The Marine Corps has focused on incremental improvements such as the Mobile Landing Platform (MLP). In reality the MLP is a less expensive solution driven by the perennial

Navy–Marine Corps debate on funding amphibious warships rather than surface combatants, the details of which we need not cover here. It is sufficient to say that the sea-basing concept deserves consideration in terms of its contribution to counter–anti-access operations, rather than simply as an enabler of amphibious operations—not currently a major Pentagon/joint priority.

A Caveat: Global Strikes from the United States May Not Be a Good Thing

Another possible alternative to a dependence on regional basing is the concept of Prompt Global Strike (PGS). In its essence, the PGS concept is that of putting conventional high explosives on a target anywhere on the globe within minutes. The idea originated in the war on al-Qaeda terrorists in Afghanistan. Intelligence was gained on Osama bin Laden's location, but available strike assets would have taken too much time to arrive at the target. Long-range bombers would take hours to fly from the United States or the U.S. base on Diego Garcia in the Indian Ocean. Tomahawk cruise missile–equipped submarines and warships were on station, but the missile is subsonic.

The proposed solution is to put conventional warheads on intercontinental ballistic missiles. Submarines would be the most likely launch platforms, but proposals for weapons launched from the continental United States continues to be discussed as an option.[6] There are concerns—particularly in think tanks and Congress—about the potential for such launches to be interpreted as nuclear attacks.[7] Those legitimate concerns aside, conducting counter–anti-access strikes from the United States shifts the focus of the offensive components of anti-access networks from attacks on regional bases to strikes on the U.S. homeland—not a positive long-range development. Although the submarine option may prove—in arsenal ship style—a leading edge breaker of anti-access networks, the continental option could take away the sanctuary from conventional war that the U.S. population has long enjoyed.

Another method of prompt global strike could be the stationing of armed unmanned aerial vehicles—drones—in the skies over or near a region of crisis or area of interest. Its counterpart would be the stationing of armed unmanned underwater vehicles (UUVs) just off the coast.

Drones and UUVs can maintain station (loiter in a designated operating area) for relatively long periods of time on relatively low amounts of fuel. Drones are subject to sovereignty restrictions, although they are stealthy enough to operate with impunity over poorly monitored areas in order to provide prompt strikes on command. UUVs would not necessarily face sovereignty restrictions (depending on where they were stationed) but might be too far away for a prompt strike on a transitory inland target. In any event, deployed drones might be a method of ensuring prompt strikes without shifting an opponent's focus to CONUS. There would be plenty of targets to shoot at in the region. On the other hand, the drones or UUVs have to be controlled from somewhere, possibly CONUS. Applying the "shoot the archer" principle, it might be tempting for the anti-access force to strike at the controllers. The ultimate point is that the consequences of adopting a PGS option as the core of a counter–anti-access force must be carefully examined in terms of where it draws the enemy's focus and whether that might not widen a regional war into a global one.

From Anti-Access to Regional War: The Anti-Access Effort Is Defeated—Now What?

Returning to a repeated complaint, America has frequently and vociferously been accused of winning the war but losing the peace. In the same vein, it may be possible to defeat an anti-access network but not thwart the anti-access state's primary objective. As previously explained, defeating an anti-access strategy is largely an enabler, in the same way that establishing sea control is an enabler of operations on land. In the language of the *JOAC*, the goal is assured access. Yet it is often an interim goal. Once access is assured, *decisive operations* on land could still have to be conducted. The modern focus on countering anti-access comes from the realization that without such a capability, decisive operations on land *cannot* be conducted. Also, a substantial (and obvious) counter–anti-access capability can be an effective deterrent to aggressive primary objectives.

It is possible, however, that a successful counter–anti-access campaign can result in the termination of hostilities if: (1) the objective of the anti-access state has not been achieved by the time the anti-access network collapses, (2) penetration of the anti-access network convinces the anti-access state that it must reach an accommodation with the strategically superior power, or (3) penetration of the anti-access network provides an opportunity for internal forces opposed to the existing regime—those within the walls—to attempt its removal. All are indeed possibilities, and they are logical reactions, but they cannot be assumed. In many cases, defeat of the anti-access network would simply be the prerequisite for continued operations on the ground, though they would very likely be successful given the elimination of anti-access capabilities. Without a counter–anti-access campaign, a regional war cannot be fought. But it most likely has to be fought to be won.

Finale: Return to Ancient Wisdom

This book began with the claim that anti-access is an ancient concept. It has analyzed anti-access strategies in terms of fundamental factors that emphasize strategic superiority, geographic characteristics, maritime predominance, criticality of information, intelligence and deception, and the impact of extrinsic events on an anti-access/counter–anti-access conflict. It has suggested likely scenarios, likely opponents, likely campaign tactics, likely system types, and issues that need to be analyzed in order to develop practical recommendations for policy. It has argued for a grand strategic approach.

It has also argued that building such walls are natural activities for authoritarian states. The book has touched upon the idea that the anti-access concept has applicability to understanding walls built to surround civil society—stretching the concept to include cognitive anti-access. However, that discussion has been left underdeveloped and unfinished. If the concept is applied to civil walls, it naturally leads to the question of whether the most effective way to collapse a military anti-access network is from within the anti-access state itself.

Demonizing or not, it has pointed to the People's Liberation Army as developing the most formidable anti-access network, even if it does not use the term. The discussion concerning *breaking the great walls* might seem provocative, but it is appropriate.

With that in mind, we end with a last bit of ancient wisdom from the legendary, anonymously authored *Thirty-Six Stratagems*. Compilations and translations may differ, yet they all start with the same stratagem: *Cross the sea without heaven's knowledge.* In other words, blind the enemy's sensors and means of intelligence while moving naval and joint forces into the region. As a first step in breaking the walls of anti-access in a world of sensors and information, there is no better advice—and of all places, from China.

Notes

Chapter 1. A Tale of Two Wars

1. Barry Strauss, *The Battle of Salamis* (New York: Simon & Schuster, 2004), 95, citing a Persian tablet. It has long been debated whether Xerxes (also known as Xerxes I) is the Persian king and consort of Esther in the biblical Book of Esther. Many theologians have concluded that his son Artaxerxes or the later king Artaxerxes II are the more likely candidates to be the biblical Xerxes.
2. Ibid., 41–42. Strauss concludes that while the number of troops appears unsupportable, 1,327 ships is a plausible number, particularly as Xerxes requisitioned the maritime forces of vassal states such as Phoenicia, Egypt, Caria, and the Ionian (Asian) Greek city-states.
3. Plutarch, *The Lives of the Noble Grecians and Romans*, trans. John Dryden, rev. Arthur Hugh Clough (New York: Modern Library, orig. 1864), 137.
4. Strauss, *Battle of Salamis*, 231–32.
5. It should be noted, however, that the Greek forces traveled most of the distance to Tempe by ship, a much faster method and one they used routinely for operational maneuver. See Herodotus, *The Histories*, trans. George Rawlinson (New York: Knopf, 1997), 581.
6. As Strauss puts it, "Tempe had been a failure of intelligence—a sign about how little the Greeks knew about their own country and how much darkness ancient strategists often worked in." Strauss, *Battle of Salamis*, 15.
7. Plutarch, *Lives*, 137.

8. This description follows Herodotus, *Histories*, 599–602. Strauss suggests that Leonides might have been planning a fighting retreat (*Battle of Salamis*, 34); most other authorities assume that his intent all along was to fight to the death.
9. Strauss, *Battle of Salamis*, 35.
10. But—without a Greek fleet to defend the sea flanks—the wall at the isthmus could be bypassed by the Persian fleet, a fact that was readily apparent to all but the most narrow-thinking of the Greek leaders. Since the Athenians and Aeginetans—from city and island, respectively, north of the isthmus—made up over half of the combined Greek fleet, the Peloponnesians could not withdraw unilaterally without losing considerable naval capability.
11. A very succinct summary of the debate over the oracle and Themistocles' actions in persuading the Athenians can be found in Sam J. Tangredi, *Futures of War: Toward a Consensus View of the Future Security Environment, 2010–2035* (Newport, RI: Alidade, 2008), 13–15. Strauss, *Battle of Salamis*, 73–123, is the best modern source for Themistocles' actions at arranging the Battle of Salamis.
12. The Battle of Salamis is traditionally highlighted as such in courses in naval history at both the U.S. Naval Academy and the U.S. Naval War College. A representative text is E. B. Potter and Chester W. Nimitz, *Sea Power: A Naval History* (Englewood Cliffs, NJ: Prentice Hall, 1960), 5–10. Potter and Nimitz's estimates for troops and ships differ from my account.
13. See the discussion in Phillip de Souza, *The Greek and Persian Wars 499–386 BC* (Oxford: Osprey, 2003), 44–46.
14. The number of 20,000 Persian troops at Plataea reflects the modern consensus that Xerxes arrived in Greece with about 150,000 to 200,000 troops total. Herodotus states that 300,000 Persian warriors were present at Plataea, but this is proportional to his overall 1.7 million total figure for the start of the Persian campaign. In any event, both series of estimates assume that Xerxes decided to withdraw around 80 percent of his force from Greece following his defeat at Salamis, which would make Salamis, not Plataea, the decisive battle of the Greek anti-access campaign. This seems in accord with what the Greeks thought at the time, based on the writings of Herodotus and the other ancient historians.
15. The Greeks used body armor more extensively, but they faced a multinational force of peoples under Persian rule with a wider variety of weapons than they possessed. Persian strategic superiority lay in numbers.
16. This is not an argument for geographic determinism or pervasive influence along the lines of Robert D. Kaplan, *The Revenge of Geography* (New York: Random House, 2012), but it corresponds with the overall importance Kaplan places on geography.

17. One region in which geography appears to be changing rapidly is the Arctic, where the sea ice mass has been steadily shrinking for the past decade. One of the first sources to discuss possible resulting issues is Jesse C. Carman, "Economic and Strategic Implications of Ice-Free Arctic Seas," in *Globalization and Maritime Power*, ed. Sam J. Tangredi (Washington, DC: National Defense University Press, 2002), 171–87.
18. A concise discussion of Poland's geopolitical dilemma is George Friedman, "Poland's Strategy," *Stratfor*, 28 August 2012, www.stratfor.com.
19. Despite not being included in the official Department of Defense dictionary (Joint Publication 1-02), *domain* has become a military planning term indicating a particular dimension of operations, generally consisting of five: land, maritime, air, space, and cyber. Ostensibly the division of operating spaces into separate domains allows for a better understanding of the forces and doctrine to be applied, with the eventual goal of *cross-domain synergy*. However, that term is burdened by connotations connected with interservice rivalry and doctrinal disputes. Specifically, U.S. Air Force doctrine holds that all joint operations within the *air domain* should be subject to an air combat commander—generally a U.S. Air Force officer—including air operations over ocean. Contrariwise, the air space over the high seas, oceanic commons, and littorals has always been seen by naval strategists as an indivisible dimension of naval warfare, with naval forces performing missions above and below the surface of the water, as well as on the surface. Indeed, naval forces provide the dominant air assets in that dimension. The U.S. Navy is therefore resistant to a definition of the *maritime domain* that includes the water but not the air above. With continuous usage of the term *domain* in the joint arena, Navy resistance has worn down a bit; however, the term maritime domain is still used differently by the Navy and Marine Corps than by the Air Force (and possibly the Army). This is the likely reason for no official definition. Throughout this book, the term maritime domain will conform to the traditional naval usage of including the air space above the high seas and littorals.
20. See the depiction of the discussions between Xerxes and the exiled Spartan former king Demaratus in Strauss, *Battle of Salamis*, 48–51.
21. On Themistocles' deceptive message to Xerxes from Salamis, see Strauss, *Battle of Salamis*, 113–18.
22. Mycale was also a battle on land. The Persians, refusing a battle at sea, had pulled up their remaining ships and built a palisade around them that was taken and burned by the marines of a Spartan-led Greek navy. The Athenian fleet did not fight at Mycale, because they had briefly evacuated their city for Salamis a second time immediately prior to the battle at Plataea.
23. Over sixty years later, the Persians gained influence across the Aegean and a degree of revenge on the Athenians by providing funds to the

Spartans in the Greek civil war against Athens and its allies known as the Peloponnesian War.

24. See Bryan Ranft and Geoffrey Till, *The Sea in Soviet Strategy* (Annapolis, MD: Naval Institute Press, 1983), 142–52, which utilizes S. G. Gorshkov, *The Sea Power of the State* (Annapolis, MD: Naval Institute Press, 1979) as a definitive source.
25. Alistair Finlan, *The Gulf War 1991* (Oxford: Osprey, 2003), 10.
26. Lawrence Freedman and Efraim Karsh, *The Gulf Conflict 1990–1991: Diplomacy and War in the New World Order* (Princeton, NJ: Princeton University Press, 1930), 176.
27. The number of Iraqi casualties in the Gulf War is still under debate, with estimates ranging from as high as 100,000 killed (acknowledged by the source as having an error factor of 50 percent) to as low as 1,000. A useful summary of the alternate estimates is the entry in Harry G. Summers, *Persian Gulf War Almanac* (New York: Facts on File, 1995), 90–91.
28. Ibid., 64, 67, 132.
29. The extent of use of space assets prompted at least one source to describe Operation Desert Storm as "the first space war." See Steven J. Bruger, "Not Ready for the First Space War: What about the Second?" *Naval War College Review* 48, no. 1 (Winter 1995), 73–83.
30. John F. Burns, "Confrontation in the Gulf: Behind Iraq's Resolve, a Militia of the Irregular," *New York Times*, 3 October 1990, www.nytimes.com/1990/10/30/world/confrontation-in-the-gulf-behind-iraq-s-resolve-a-militia-of-the-irregular.html (accessed 1 November 2012).
31. The Correlates of War project was started in 1963 by political scientist J. David Singer at the University of Michigan as a quantitative research effort to determine the causes of war. It maintains a website at www.correlatesofwar.org. It was partly inspired by earlier research by the British statistician and meteorologist Lewis Fry Richardson as exemplified in his book *Statistics of Deadly Quarrels* (Pittsburgh: Boxwood, 1960). The raw data from these sources, particularly the number of forces involved and casualties of historical conflicts, have proven useful to many researchers, and these efforts have had a considerable impact on the quantitative approach to political science research. However, it must be kept in mind that they were created with a foundational bias toward the belief that institutional forces are the likely cause of war and that war as an activity can eventually be eliminated.
32. U.S. Department of Defense, *Joint Operational Access Concept*, Version 1.0 (Washington, DC: Department of Defense, 2012), ii.
33. General K. Sundjari was Indian Army chief of staff from 1986 to 1988 and died in 1999. Although I have been unable to trace its original source, versions of his quote have been used or referred to in a variety of assessments, the latest being Elbridge Colby, "The Role of Nuclear Weapons in the Global Security Environment," *Global Trends 2030*

(blog), 29 June 2012, http://gt20230.com/2012/06/29/the-role-of-nuclear-weapons-in-the- future security-environment-2/ (accessed 1 November 2012). An early citation is in Robert G. Joseph and John F. Reichart, *Deterrence and Defense in a Nuclear, Biological, and Chemical Environment* (Washington, DC: National Defense University, Institute for National Strategic Studies, 1995), 4.

34. From "Silver Blaze" in Arthur Conan Doyle, *Memoirs of Sherlock Holmes*, various editions.
35. Gilberto Villahermosa, *Desert Storm: The Soviet View* (Ft. Leavenworth, KS: Foreign Military Studies Office, 1992), 2, http://fmso.leavenworth.army.mil/documents/res-storm (accessed 1 November 2012).
36. Michael Pilsbury, *China Debates the Future Security Environment* (Washington, DC: National Defense University Press, 2000), 103.
37. Paul H. B. Godwin, "PLA Doctrine and Strategy: Mutual Apprehension in Sino-American Military Planning," in *The People's Liberation Army and China in Transition*, ed. Stephen J. Flanagan and Michael E. Marti (Washington, DC: National Defense University Press, 2003), 268.
38. U.S. Department of Defense, *Conduct of the Persian Gulf War: Final Report to Congress* (Washington, DC: Department of Defense, 1992), 23, http://www.ndu.edu/library/epubs/cpgw.pdf (accessed 1 November 2012).
39. Freedman and Karsh, *Gulf Conflict*, 176.
40. Ibid., 441.
41. As Paul Bracken puts it, "international condemnation forced those who wanted the bomb to go underground. Desiring an atomic bomb—for those countries that didn't already have one—became viewed as a kind of perversion, something countries did, if at all, in deepest secrecy." Bracken, *Fire in the East: The Rise of Asian Military Power and the Second Nuclear Age* (New York: HarperCollins, 1999), 99.
42. Ibid., 45–48.
43. National Defense Panel, *Transforming Defense: National Security in the 21st Century* (Arlington, VA: National Defense Panel, 1997), i.
44. Roger Cliff, Mark Burles, Michael S. Chase, Derek Eaton, and Kevin L. Pollpeter, *Entering the Dragon's Lair: Chinese Antiaccess Strategies and Their Implications for the United States* (Santa Monica, CA: RAND Corp., 2007), 2.

Chapter 2. Developing the Modern Concept of Anti-Access

1. Despite the president being commander in chief of the armed forces, it is relatively unusual for him to sign a stand-alone document giving priorities to the Department of Defense. Usually this is done through the president's signature on the National Security Council staff–produced *National Security Strategy of the United States* (*NSS*). The *NSS* is still legally required as an annual report to Congress despite the fact that recent

presidential administrations have issued it on a less regular basis, with President George W. Bush issuing only one during his term. President Obama's administration has also issued one, in 2010. Normally separate defense-priority documents are issued in the name of the secretary of defense, and indeed Secretary Leon Panetta also signed the document with his own cover letter. The fact that the presidential signature is unusual was noted in a *Foreign Policy* magazine blog authored by Peter Feaver: "The Administration went to considerable lengths to emphasize the boldness and novelty of what they were they were doing." Posted 3 February 2012 at shadow.foreignpolicy.com/posts/2012/02/03/should_the_obama_administration_relase_another_national_security_strategy_this_ter (accessed 1 September 2012).

2. U.S. Department of Defense, *Sustaining U.S. Global Leadership: Priorities for 21st Century Defense* (Washington, DC: Office of the Secretary of Defense, 2012), 4.
3. Ibid., 4–5.
4. U.S. Department of Defense, *Joint Operational Access Concept*, 6.
5. Ibid.
6. Ibid., i.
7. Ibid., i, 6.
8. Ibid., 1.
9. Ibid.
10. Ibid.
11. Walter Raleigh, *The Works of Walter Raleigh, Kt.*, vol. 8 (Oxford, UK: Oxford University Press, 1829), 325.
12. Alfred Thayer Mahan, *Lessons of the War with Spain and Other Articles* (Boston: Little, Brown, 1899), 106.
13. U.S. Department of Defense, *Joint Operational Access Concept*, 3.
14. The term *sea control* was particularly popularized in the mid-1970s by Vice Adm. Stansfield Turner, USN, during his tenure as president of the U.S. Naval War College. See Stansfield Turner, "Missions of the U.S. Navy," *Naval War College Review* 26, no. 5, March–April 1974, 2–17.
15. This derivation is evidenced in Andrew F. Krepinevich Jr., *The Military-Technical Revolution: A Preliminary Assessment* (Washington, DC: Center for Strategic and Budgetary Assessments, 2002), which reproduces a study originally written in 1992. Krepinevich applies the denial concept to other mediums, noting (in the case of space) that "non-peer competitors will probably set space denial as their objective, much in the manner in which inferior naval powers have traditionally sought sea denial as opposed to sea control" (pp. 23–24).
16. Issued as a supplement to U.S. Naval Institute *Proceedings*, January 1986. For extensive discussion of the Maritime Strategy, see Norman

Friedman, *The U.S. Maritime Strategy* (Annapolis, MD: Naval Institute Press, 1988).

17. A profile of Marshall's influence is Jay Winik, "Secret Weapon: The Pentagon's Andy Marshall Is the Most Influential Man You've Never Heard Of," *Washingtonian*, April 1999, 45–55.
18. Curiously the "anti-Navy" term has been revived in describing Chinese anti-access developments, most likely in ignorance of its earlier usage. See Andrew S. Erickson, "Are China's Near Seas "Anti-Navy" capabilities aimed directly at the United States?," *Information Dissemination* blog, 14 June 2012, www.informationdissemination.net/2012/06/are-chinas-near-seas-anti-navy.htm (accessed 1 August 2012).
19. U.S. Navy, Office of Naval Intelligence, *Challenges to Naval Expeditionary Warfare* (Washington, DC: Office of Naval Intelligence, 1997).
20. Sam J. Tangredi, ed., *Globalization and Maritime Power* (Washington, DC: National Defense University Press, 2002).
21. Quoted in Adam B. Siegel, "Base Access Constraints and Crisis Response," *Airpower Journal* 10, no. 2, Spring 1996, 107.
22. Ibid., note 1.
23. One of the most influential articles was Andrew Krepinevich Jr., "Cavalry to Computer: The Pattern of Military Revolutions," *National Interest*, no. 37 (Fall 1994), 30–42.
24. John P. Jumper, "Expeditionary Air Force: A New Culture for a New Century," Air Force Association, 26 February 1998, www.afa.org/afe/pub/ol19.asp (accessed 1 July 2012).
25. Quoted in Cliff et al., *Entering the Dragon's Lair*, 1.
26. The quote is from ibid., 1, referencing the Defense Science Board, *Report of the Defense Science Board Task Force on Strategic Mobility* (Washington DC: Government Printing Office, 1996), 52–56.
27. U.S. Secretary of Defense, *Quadrennial Defense Review* (Washington, DC: Office of the Secretary of Defense, 1997), 4.
28. National Defense Panel, *Transforming Defense*, 12.
29. Ibid., 13.
30. See Thomas G. Mahnken, "Deny U.S. Access?," U.S. Naval Institute *Proceedings* 124, no. 9 (September 1998), 36–39.
31. Kenneth F. McKenzie Jr., *The Revenge of the Melians: Asymmetric Threats and the Next QDR*, McNair Paper 62 (Washington, DC: National Defense University Press, 2000); Sam J. Tangredi, *All Possible Wars? Toward a Consensus View of the Future Security Environment, 2001–2025*, McNair Paper 63 (Washington, DC: National Defense University Press 2000).
32. Clark A. Murdock, "The Navy in an Antiaccess World," 473–86, and Norman Friedman, "Globalization of Antiaccess Strategies," 487–502, in Tangredi, ed., *Globalization and Maritime Power.*

33. Paul K. Davis, Jimmie McEver, and Barry Wilson, *Measuring Interdiction Capabilities in the Presence of Anti-Access Strategies: Exploratory Analysis to Inform Adaptive Strategy for the Persian Gulf* (Santa Monica, CA: RAND Corp., 2002), xiii.
34. Ibid., 1.
35. Andrew Krepinevich, *Why AirSea Battle?* (Washington, DC: Center for Strategic and Budgetary Assessments, 2010), 13–14.
36. Cliff et al., *Entering the Dragon's Lair*, 28.
37. Ibid., 27.
38. See for example, Jan Perlez, "Singaporean Tells China U.S. Is Not in Decline," *New York Times*, 6 September 2012, www.nytimes.com/2012/09/07world/asia/singapores-prime-minister-warns-china-on-view-of-US.htm (accessed 1 November 2012).
39. Christopher J. Bowie, *The Anti-Access Threat to Theater Air Bases* (Washington, DC: Center for Strategic and Budgetary Assessments, 2002), iii.
40. Ibid., i.
41. Ibid.
42. Robert O. Work, *The Challenge of Maritime Transformation: Is Bigger Better?* (Washington, DC: Center for Strategic and Budgetary Assessments, 2002).
43. Andrew Krepinevich, Barry Watts, and Robert Work, *Meeting the Anti-Access and Area Denial Challenge* (Washington, DC: Center for Strategic and Budgetary Assessments, 2003), i.
44. Ibid.
45. Ibid., iv.
46. U.S. Department of Defense, *Joint Operational Access Concept*, 4.
47. Ibid.
48. Norton A. Schwartz and Jonathan W. Greenert, "Air-Sea Battle: Promoting Stability in an Era of Uncertainty," *The American Interest*, 20 February 2012, www.the-american-upinterest.com/article.cfm?piece=1212 (accessed 1 April 2012).
49. Sam LaGrone, "Pentagon's 'Air-Sea Battle' Plan Explained. Finally," *Wired.com Danger Room*, 6 August 2012, www.wired.com/dangerroom/2012/08/air-sea-battle-2/ (accessed 1 September 2021).
50. Phillip Ewing, "The Rise and Fall of Air-Sea Battle," *DoD Buzz: Online Defense and Acquisition Journal*, 1 May 2012, www.dodbuzz.com/2012/05/17/the-rise-and-fall-of-air-sea-battle/ (accessed 1 August 2012).
51. Ibid.
52. LaGrone.

53. Krepinevich, *Why AirSea Battle?*, 1.
54. Ibid., 5.
55. Ibid.
56. Ibid., 7.
57. Ibid., 114.
58. Jan Van Tol with Mark Gunzinger, Andrew Krepinevich, and Jim Thomas, *AirSea Battle: A Point of Departure Operational Concept* (Washington, DC: Center for Strategic and Budgetary Assessments, 2010).
59. Roger Barnett, *Asymmetrical Warfare: Today's Challenges to U.S. Military Power* (Washington, DC: Brassey's, 2002), 15.
60. Rod Thornton, *Asymmetric Warfare: Threat and Response in the Twenty-First Century* (London: Polity, 2007), 4.
61. Pillsbury, *China Debates*, 292.
62. Krepinevich, *Military-Technical Revolution*, 44.
63. George Friedman makes the point that it is not the general U.S. population that is casualty-averse but America's governing elite in its "vast distance" from those who fight. Friedman, *America's Secret War: Inside the Hidden Worldwide Struggle between America and Its Enemies* (New York: Doubleday, 2004), 338–39.
64. On forward presence as a strategic concept, see Sam J. Tangredi and Randall G. Bowdish, "Core of Naval Operations: Strategic and Operational Concepts of the United States Navy," *The Submarine Review*, January 1999, 14–15.
65. See discussion in Sam J. Tangredi, "The Fall and Rise of Naval Forward Presence," U.S. Naval Institute *Proceedings* 128, no. 5 (May 2000), 28–32.
66. Friedman, "Globalization of Antiaccess Strategies," 489–95.
67. Krepinevich, Watts, and Work, *Meeting the Anti-Access and Area Denial Challendge*, 50, 59–61.
68. Cliff et al., *Entering the Dragon's Lair*, 44–45, 51–60.
69. Van Tol et al., *Point of Departure*, xii.
70. A good exposition is Timothy A. Walton, *China's Three Warfares* (Washington, DC: Delex Systems, 2012).
71. John M. Collins, *Grand Strategy: Principles and Practices* (Annapolis, MD: Naval Institute Press, 1973), 14–15.
72. John A. Warden III, *The Air Campaign: Planning for Combat* (Washington, DC: National Defense University Press, 1988), 5, and "The Enemy as a System," *Airpower Journal* 9, no. 2 (Spring 1995), 40–55.

Chapter 3. The Anti-Access Campaign and Its Defeat

1. J. C. Wylie, *Strategy: A Theory of Power Control* (New Brunswick, NJ: Rutgers University Press, 1967; republished with introduction and new postscript, Annapolis, MD, Naval Institute Press, 1989), 14.
2. Dudley W. Knox, *The Naval Genius of George Washington* (Boston: Houghton Mifflin, 1932), 13.
3. Sun Tzu, *The Art of War*, trans. Samuel B. Griffith (New York: Oxford University Press, 1963), 77.
4. Quoted in Margaret MacMillan, *Nixon, Kissinger, and the Opening to China* (New York: Random House, 2008), 144.
5. Walton, *China's Three Warfares*.
6. William L. Shirer, *The Rise and Fall of the Third Reich* (New York: Simon & Schuster, 1960), 394.
7. U.S. Congress, Joint Committee on the Investigation of the Pearl Harbor Attack, *Pearl Harbor Attack*, part 12, 79th Cong., 1st sess. (Washington, DC: Government Printing Office, 1946), 195.
8. A recent discussion on Acheson's remarks is James I. Matray, "Dean Acheson's Press Club Speech Reexamined," *Journal of Conflict Studies* 22, no. 1 (Spring 2002), http://journals.hil.unb.na/index.php/jcs/article/view366/578 (accessed 1 June 2012).The details of Ambassador Glaspie's fateful meeting with Saddam are still controversial. One possibly inaccurate transcript version is "Excerpts from Iraqi Document on Meeting with U.S. Envoy," *New York Times International*, September 23, 1990, http://msuweb.montclair.edu/~furrg/glaspie.html (accessed 1 June 2012). More recent comments by Ambassador Glaspie are in Glenn Kessler, "Ex-Envoy Details Hussein Meeting," *Washington Post*, April 3, 2008, http://www.washingtonpost.com/wp-dyn/content/article/2008/04/02/AR2008040203485.html (accessed 1 June 2012).
9. U.S. Department of Defense, *Joint Operational Planning*, Joint Publication 5.0, (Washington, DC: Joint Chiefs of Staff, 2011), III-36.
10. Ibid., III-43.
11. Friedman, "Globalization of Antiaccess Strategies," 488–89.
12. Cliff et al., *Entering the Dragon's Lair*, 51–93.
13. Van Tol et al., *AirSea Battle*, 18–19.
14. Ibid.
15. Derived from chart, ibid., 18.
16. This dialogue includes: Andrew S. Erickson and David D. Yang, "Using the Land to Control the Seas: Chinese Analysts Consider the Antiship Ballistic Missile," *Naval War College Review* 62, no. 4 (Autumn 2009), 53–86; Andrew S. Erickson and David D. Yang, "On the Verge of a Game Changer," U.S. Naval Institute *Proceedings* 135, no. 5 (May 2009), 26–32; Sam J. Tangredi, "No Game Changer for China," U.S. Naval Institute

Proceedings 136, no. 2 (February 2010), 24–29; and Craig Hooper and Christopher Albon, "Get Off the Fainting Couch," U.S. Naval Institute *Proceedings* 136, no. 4 (April 2010), 42–46.

17. Friedman, "Globalization of Antiaccess Strategies," 488.
18. Quoted in Basil Liddell Hart and Adrian Liddell Hart, *The Sword and the Pen* (New York: Thomas Y. Crowell, 1976), 180.
19. Strauss, *Battle of Salamis*, 98.

Chapter 4. Three Anti-Access Victories

1. Sources consulted include Roger Hart, *Battle of the Spanish Armada* (East Sussex, UK: Wayland, 1973); Peter Kemp, *The Campaign of the Spanish Armada* (New York: Facts on File, 1988); Colin Martin and Geoffrey Parker, *The Spanish Armada* (New York: Norton, 1988); Garrett Mattingly, *The Armada* (Boston: Houghton Mifflin, 1959); Alexander McKee, *From Merciless Invaders: An Eyewitness Account of the Spanish Armada* (New York: Norton, 1963); and Jay Williams, *The Spanish Armada* (New York: American Heritage, 1966).
2. In compiling his 1963 interpretation, subtitled *An Eyewitness Account of the Spanish Armada,* British journalist Alexander McKee notes that "what impressed above all was the quality of the eye-witness narratives. Never, in the long history of England's wars, can a set of combat reports have been written to equal these. The red-hot phrases shower from the battle like sparks from a grinder. One ceases to marvel so much at Shakespeare." McKee, *From Merciless Invaders*, 9.
3. Mattingly, *Armada*, 77.
4. Ibid., 78.
5. Quoted in Kemp, *Campaign of the Spanish Armada*, 60.
6. Martin and Parker, *Spanish Armada*, 261.
7. Sources consulted include Eliot A. Cohen and John Gooch, *Military Misfortunes: The Anatomy of Failure in War* (New York: Free Press, 2005); Julian Corbett, *History of the Great War: Naval Operations*, vols. 1 and 2 (London: Longmans, Green, 1920, 1921); Theodore L. Gatchel, *At the Water's Edge: Defending against the Modern Amphibious Assault* (Annapolis, MD: Naval Institute Press, 1996); Roger Keyes, *The Fight for Gallipoli* (London: Eyre & Spottiswoode, 1934); Robert K. Massie, *Castles of Steel: Britain, Germany, and the Winning of the Great War* (New York: Random House, 2003); Alan Moorehead, *Gallipoli* (Hertfordshire, UK: Wordsworth, 1997); Phil Taylor and Pam Cupper, *Gallipoli: A Battlefield Guide* (Kenthurst, Australia: Kangaroo Press, 1989); Victor Ruddeno, *Gallipoli: Attack from the Sea* (New Haven, CT: Yale University Press, 2008); Peter Shankland, *Dardanelles Patrol* (New York: Scribner, 1964); and Dan Van Der Vat, *The Dardanelles Disaster: Winston Churchill's Greatest Failure* (New York: Overlook, 2009).

8. Van Der Vat, *Dardanelles Disaster*, ix.
9. Ibid., 108.
10. Some sources portray the report of the Turkish lack of ammunition as a legend generated by later miscalculations by British naval intelligence. But at the same time, many Turkish commanders appear to have concluded that it was only a matter of time until the British fleet made its way to Constantinople where the Turkish army would have to fight a decisive land battle. See Van Der Vat, *Dardanelles Disaster*, 129, in contrast to Moorehead, *Gallipoli*, 59–74. Moorehead reports that the "Young Turks" regime had made preparations to burn Constantinople to the ground rather than surrender it to "infidels." British submarines did penetrate into the Sea of Marmara and beyond the Bosporus where they commenced sinking Turkish ships—a situation that encouraged pessimistic estimates by the Turks.
11. Sources consulted include Stephen Bungay, *The Most Dangerous Enemy: A History of the Battle of Britain* (London: Aurum, 2000); Winston Churchill, *The Second World War, Volume 2: Their Finest Hour* (London: Cassell, 1949); Len Deighton, *Fighter: The True Story of the Battle of Britain* (New York: Knopf, 1978), and *Battle of Britain* (London: George Rainbird, 1980); Peter Fleming, *Operation Sea Lion* (New York: Simon & Schuster, 1957); Gatchel, *Water's Edge*; Martin Gilbert, *Winston S. Churchill, Volume 6: Finest Hour, 1939–1941* (London: Heinemann, 1983); James Holland, *The Battle of Britain: Five Months That Changed History; May–October 1940* (New York: St. Martin's Griffin, 2012); David Irving, *Hitler's War* (New York: Viking, 1977); Egbert Kieser, *Hitler on the Doorstep: Operation "Sea Lion"; The German Plan to Invade Britain, 1940* (Annapolis, MD: Naval Institute Press, 1997); Michael Korda, *With Wings like Eagles: A History of the Battle of Britain* (New York: HarperCollins, 2009); Richard Overy, *Battle of Britain: The Myth and Reality* (New York: Norton, 2001); and Derek Robinson, *Invasion 1940: The Truth about the Battle of Britain and What Stopped Hitler* (New York: Da Capo Press, 2005).
12. Deighton, *Fighter*, xvi.

Chapter 5. Three Anti-Access Defeats

1. Use of the term Islas Malvinas is not intended to represent support for Argentine claims to the Falkland Islands. Modern international law, which is built around the self-determination of peoples, clearly supports the Falkland Islanders' choice to remain in political association with the United Kingdom.
2. Samuel Eliot Morison, *Strategy and Compromise* (Boston: Little, Brown, 1958), 23–24.
3. Ibid., 45.
4. Gatchel, *Water's Edge*, 60.

5. Irving, *Hitler's War*, 407.
6. Gatchel, *Water's Edge*, 60.
7. Ibid., 65.
8. Ibid.
9. Ibid., 66.
10. Ibid., 69.
11. Stephen E. Ambrose, *Americans at War* (Jackson: University Press of Mississippi, 1997), 67.
12. For details on Ultra, see F. W. Winterbotham, *The Ultra Secret* (New York: Harper & Row, 1974). On general deception, see Ewen Montagu, *Beyond Top Secret Ultra* (New York: Coward, McCann & Geoghegan, 1978). Both Winterbotham and Montagu were participants.
13. Ronald Lewin, *The American Magic: Codes, Ciphers and the Defeat of Japan* (New York: Farrar, Straus and Giroux, 1982), 105.
14. The primary source used for this discussion is Edwin P. Hoyt, *Yamamoto: The Man Who Planned Pearl Harbor* (New York: McGraw-Hill, 1990), but the details appear in many other sources. Http://en.wikipedia.org/wiki/Isoroku_Yamamoto (accessed 1 June 2012) appears well researched.
15. Gatchel, *Water's Edge*, 95.
16. Ronald H. Spector, *Eagle against the Sun: The American War with Japan* (New York: Free Press, 1985), 76–77.
17. Gordon Prange, *At Dawn We Slept* (New York: Penguin, 1982), 11.
18. This quote has been reported by a number of sources as originating in a private letter between Yamamoto and Prime Minister Fumimaro Konoe in 1940. A slightly different version appears in Frederick D. Parker, *A Priceless Advantage: U.S. Navy Communications Intelligence and the Battles of Coral Sea, Midway and the Aleutians* (Washington, DC: National Security Agency, 2001), 1.
19. Lewin, *American Magic*, 99.
20. Ibid., 109.
21. Spector, *Eagle*, 81.
22. Ibid., xiv–xv, 118–19.
23. Sources consulted include Michael Clapp and Ewen Southby-Tailyour, *Amphibious Assault Falklands: The Battle of San Carlos Water* (Barnsley, UK: Pen & Sword, 2007); Norman Friedman, "The Falklands War: Lessons Learned and Mislearned," *Orbis* 26, no. 4 (Winter 1983), 907–40; Gatchel, *Water's Edge*; Douglas M. Hime, *The 1982 Falklands-Malvinas Case Study*, NWC 1036 (Newport, RI: U.S. Naval War College, 2010); Martin Middlebrook, *Operation Corporate: The Story of the Falklands War, 1982* (London: Viking, 1985), and *The Fight for the Malvinas: The Argentine Forces in the Falklands War* (London: Viking, 1989); *Sunday Times* of London Insight Team, *War in the Falklands: The Full Story* (New York: Harper & Row, 1982); Harry

D. Train III, "An Analysis of the Falkland/Malvinas Islands Campaign," *Naval War College Review* 41, no. 2 (Winter 1988), 33–50; and John Woodward with Patrick Robinson, *One Hundred Days: The Memoirs of the Falklands Battle Group Commander* (Annapolis, MD: Naval Institute Press, 1992). Wikipedia articles on the Falklands War appear well researched.

24. Christopher Chant, *The Military History of the United States: Revolutionary and Early American Wars* (New York: Webster's Home Library, 1997), 195.
25. Middlebrook, *Operation Corporate*, 35.
26. Ibid., 35–36.
27. Hime, *Falklands-Malvinas Case Study*, 17.
28. Ibid., 58.
29. Middlebrook, *Operation Corporate*, 142–52.
30. Jean Labaye Couhat and A. D. Baker, eds., *Combat Fleets of the World 1986/87* (Annapolis, MD: Naval Institute Press, 1986), 6; Gatchel, *Water's Edge*, 195.
31. Gatchel, *Water's Edge*, 195.
32. Middlebrook, *Operation Corporate*, 385.
33. http://en.wikipedia/wiki/Argentine_ground_forces_in_the_Falklands_War, 5 (accessed 1 July 2012). Although Wikipedia is viewed with suspicion as a reliable source, this appears well referenced.
34. Middlebrook, *Operation Corporate*, 76.
35. Ibid., 148–49.
36. *Sunday Times* of London Insight Team, *War in the Falklands*, 45–57.
37. Specter, *Eagle*, 80.

Chapter 6. East Asia

1. Quoted in Robert D. Heinl, *Dictionary of Military and Naval Quotations* (Annapolis, MD: Naval Institute Press, 1966), 239. However, other sources indicate that the correct quotation reads: "No plan of operations extends with certainty beyond the first encounter with the enemy's main strength."
2. Richard Danzig, *The Big Three: Our Greatest Security Risks and How to Address Them* (New York: Center for International Political Economy, 1999), 6.
3. For a discussion on consensus assessment of the future, representative source comparative analysis, and the use and limits of forecasts and scenarios, see Tangredi, *Futures of War*, especially 43–59.
4. Ewing, *Rise and Fall*; Sydney J. Freedberg, "Cartwright Targets F-35, AirSea Battle: Warns of $250B More Cuts," *AOL Defense*, 15 May

2012, http://defense.aol.com/2012/05/15/cartwright-savages-f-35-airsea-battle-warns-of-250-billion-mo/ (accessed 1 June 2012).

5. It might strike readers with an appreciation for the ironies of history that remarks about the need to avoid "demonizing" China have unintended parallels to concerns in the 1930s to avoid "demonizing" Germany, lest Hitler be so upset that he do something unpleasant. This is not intended to suggest that the collective CCP would act like Hitler, only that the perspective that "demonizing" China through discussion or planning for anti-access challenges seems to imply that it is the United States, not the CCP, that controls China's destiny.
6. Although dated in terms of the newest developments, an excellent brief explanation of maritime satellite surveillance is Arthur H. Barber III and Delwyn L. Gilmore, "Maritime Access: Do Defenders Hold All the Cards?" *Defense Horizons*, no. 4 (October 2001), 4–5.
7. On the view of China as "hemmed in" by the island chains, see Gabriel B. Collins, "China's Dependence on the Global Maritime Commons," in *China, the United States and 21st-Century Sea Power: Defining a Maritime Security Partnership*, ed. Andrew S. Erickson, Lyle Goldstein, and Nan Li (Annapolis, MD: Naval Institute Press, 2010), 20–23.
8. Bernard D. Cole, *The Great Wall at Sea, Second Edition: China's Navy in the Twenty-First Century* (Annapolis, MD: Naval Institute Press, 2010).
9. These scholars include Andrew S. Erickson, Lyle J. Goldstein, James R. Holmes, Nan Li, and Toshi Yoshihara. Andrew Erickson runs excellent blogs that provide updates on naval developments and PRC news, one individually at www.andrewserickson.com and one with Gabriel B. Collins at China Signpost (www.chinasignpost.com). James R. Holmes also has a regular blog on the PLAN and other naval topics (http://thediplomat.com/the-naval-diplomat).
10. Voice of Russia, "China Interested in Russian Missile System," 10 May 2012, Space War, http://spacewar.com/reports/China_interested_in_Russian_missile_system-999.html (accessed 1 June 2012).
11. IHS Jane's, "Samsung-Thales Begins M-SAM Radar Production," *Jane's Missiles and Rockets* (London: Jane's Defence Group, 2007), posted at http://www.janes.com/articles/Janes-Missiles-And-Rockets-2007/Samsung-Thales-begins-M-SAM-radar-production.html (accessed 1 August 2012).
12. Upi.com, "Riyadh Mulls Big Russian Missile Buy," 22 March 2010, www.upi/Business_News/Security-Industry/2010/03/22/Riyadh-mulls-big-Russian-missile-buy/UPI-9311269283852 (accessed 1 August 2012).
13. IHS Jane's, "First Photos of S-400 in China," *Jane's Strategic Weapon Systems* (London: Jane's Defence Group, 2008), posted on China Defense Blog, http//china-defense.blogspot.com/2009/05/first-photos-of-s-400-in-china.html.

14. Erickson and Yang, "Using the Land," 53–86, and "On the Verge," 26–32.
15. Erickson and Yang, "Using the Land," 56.
16. See Tangredi, "No Game Changer," 24–29, and Hooper and Albon, "Get off the Fainting Couch,"42–46.
17. Richard C. Bush and Michael E. O'Hanlon, *A War like No Other: The Truth about China's Challenge to America* (Hoboken, NJ: John Wiley, 2007).
18. Thomas P. M. Barnett, *The Pentagon's New Map: War and Peace in the Twenty-First Century* (New York: Berkley, 2004), 242.
19. Ralph Peters, *Fighting for the Future: Will America Triumph?* (Mechanicsburg, PA: Stackpole, 1999), 223.
20. For an excellent brief introduction to HEMP, see McKenzie, *Revenge of the Melians*, 34–39.
21. A short introductory article on nonnuclear EMP weapons that appeared in a general news periodical is "Frying Tonight: Warfare Is Changing as Weapons That Destroy Electronics, Not People, Are Deployed on the Field of Battle," *The Economist*, 15 October 2011, posted at http://www.economist.com/node/21532245 (accessed 1 July 2012).
22. Global Security.org, "China Building EMP Arms in Case of Conflict over Taiwan: Report," July 24, 2011, http://www.globalsecurity.org/wmd/library/news/china/2011/china-110724-cna01.htm (accessed 1 July 2012).
23. This is a theme that runs throughout Hughes' work. See for example Wayne P. Hughes Jr., *Fleet Tactics and Coastal Combat*, 2nd ed. (Annapolis, MD: Naval Institute Press, 2000), 40–44.

Chapter 7. Southwest Asia

1. This argues for the operation of naval forces in a widely dispersed disposition, one of the premises of the once popular concept of network-centric warfare.
2. Jane's ongoing series of annual or biennial weapons publications such as *Jane's Fighting Ships* (Stephen Saunders, ed.), *Jane's All the World's Aircraft* (Paul Jackson, ed.), and *Jane's Land-Based Air Defence* (James C. O'Halloran, ed.), all published in London by IHS Jane's Information Group; International Institute for Strategic Studies, *The Military Balance 2012* (London: Routledge, 2012); Stockholm International Peace Research Institute, *SIPRI Yearbook 2012* (New York: Oxford University Press, 2012); U.S. Department of Defense, *Annual Report on Military Power of Iran* (Washington, DC: Office of the Secretary of Defense, 2012).
3. Mark Gunzinger, "Outside-In: Defeating Iran's Anti-Access and Area-Denial Threat" (briefing slides), CSBA, 17 January 2012, 4.
4. Ibid.

5. The fact that the Islamic Republic of Iran maintains the trappings of elections hardly betrays the fact that the theocratic autocracy manipulates and controls the outcome. Hooman Majd—who appears to believe that a theocratic democracy is a possibility—details this control in his book *The Ayatollahs' Democracy: An Iranian Challenge* (New York: Norton, 2010).
6. The 1983 Beirut bombing that killed 241 U.S. servicemen was carried out by Hezbollah, with an Iranian minister taking public credit for providing the explosives. See Alireza Jafarzadeh, *The Iran Threat: President Ahmadinejad and the Coming Nuclear Crisis* (New York: Palgrave Macmillan, 2007), 66–67.
7. Many analysts maintain that the cooperation between Sunni and Shiite terrorist groups and Iran is greater than recognized. See, for example, Ronen Bergman, *The Secret War with Iran: The 30-Year Clandestine Struggle against the World's Most Dangerous Terrorist Power* (New York: Free Press, 2008), 213–35.
8. Former FBI director Louis Freeh maintains that the 1996 attack of the Khobar Towers in Saudi Arabia, in which nineteen U.S. service members were killed, was planned and directed by the Iranian government. See Jafarzadeh, *Iran Threat*, 72–73. In 2006, an Arabic newspaper in the United Kingdom quoted a senior Iranian official describing Hezbollah in Lebanon as "one of the foundations of our security strategy. It serves as the first Iranian defensive line against Israel." Bergman, *Secret War*, 254.
9. Judith S. Yaphe and Charles D. Lutes, *Reassessing the Implications of a Nuclear-Armed Iran*, McNair Paper 69 (Washington, DC: National Defense University Press, 2005), 12–13. Yaphe and Lutes maintain that "we know little about how Iranian leaders make their decisions, what factors influence them, or what understanding they have of U.S., E.U., or Israeli 'redlines' regarding actions Iran might take."
10. Kenneth Waltz, "Why Iran Should Get the Bomb," *Foreign Affairs* 91, no. 4 (July/August 2012), 2–5. Waltz's argument assumes that Cold War–style deterrence based on the logic of game theory will always work. As previously discussed, that remains an unproven assumption.
11. Amir Taheri maintains that the founding supreme leader of the Islamic Republic, Ayatollah Khomeini, "described war as a divine blessing." Taheri also makes the point that none of the other current authoritarian regimes has had the same "messianic pretensions or seen itself as in the vanguard of global conquest in the name of religion." Amir Taheri, *The Persian Night: Iran under the Khomeinist Revolution* (New York: Encounter, 2009), 95, 311. Concerning wars of religion, the writings of Ralph Peters are both enlightening and alarming. See Peters, *Endless War: Middle Eastern Islam vs. Western Civilization* (Mechanicsburg, PA: Stackpole Books, 2010).

12. Mark Gunzinger with Chris Dougherty, *Outside-In: Operating from Range to Defeat Iran's Anti-Access and Area Denial Threats* (Washington, DC: CSBA, 2011), x.
13. The fascination with swarming seems to assume that large numbers of attackers can be brought to close quarters with individual defenders, not necessarily an easy prospect against alerted defenders operating as a networked force.
14. Gunzinger with Dougherty, *Outside-In*, 22–23.
15. Ibid., 34–38.
16. Jafarzadeh, *Iran Threat*, 54–56, 207–9.
17. Yaphe and Lutes, *Reassessing the Implications*, 1–8.
18. Majd quotes the theological autocracy–controlled daily newspaper *Kayan* as stating: "In the power struggle in the Middle East, there are only two sides: Iran and the United States." Majd, *Ayatollahs' Democracy*, 132.
19. Kenneth Katzman, Neelesh Nerurkar, Ronald O'Rourke, R. Chuck Mason, and Michael Ratner, *Iran's Threat to the Strait of Hormuz*, Report for Congress R42335 (Washington, DC: Congressional Research Service, 2012), 1.
20. Ibid., 15–16.
21. Taheri, *Persian Night*, 149.
22. Majd provides extensive details, but denies that it was a revolution. Rather, it was a reaction to the blatant control of the presidential election by the theocratic autocracy. Majd, *Ayatollah's Democracy*, 52–60.
23. Representative discussions of nuclear weapons use in asymmetrical warfare: McKenzie, *Revenge of the Melians*, 20–26; U.S. Department of Defense, *Military Power of Iran*, 4.
24. Bracken argues that a "second nuclear age" would be a very natural consequence. Bracken, *Fire in the East*, 95–124.
25. Sidney J. Freedberg, "Iran Mine Threat Scares Navy; CNO Scrambles to Fix Decades of Neglect," *AOL Defense*, 4 May 2012, http://defense.aol.com/2012/05/04/iran-mine-threat-scares-navy-cno-scrambles-to-fix-decades-of-neglect.html (accessed 1 September 2012); Nick Schiffrin and Matthew McGarry, "Iran, US Flex Military Muscles in Persian Gulf: US Carries Out Vast Allied Anti-Mine Exercise, Iran Test Fires Anti-Warship Missiles," ABC News, 25 September 2012, abcnews.go.com (accessed 1 September 2012).
26. The conclusion of Robert E. Looney, "Market Effects of Naval Presence in a Globalized World: A Research Summary," in *Globalization and Maritime Power*, ed. Tangredi, 103–31.
27. McKenzie, *Revenge of the Melians*, 14–16.
28. Clifford Krause, "Oil Price Would Skyrocket if Iran Closed the Strait of Hormuz," *New York Times*, 4 January 2012, http://www.nytimes

.com/2012/01/05/business/oil-price-would-skyrocket-if-iran-closed-the-strait-of-hormuz/ (accessed 1 September 2012).

29. The Wikipedia article on Stuxnet at http://en.wikipedia.org/wiki/Stuxnet (accessed 1 October 2012) appears well researched and includes some speculation that the attack on Iranian centrifuges was actually designed as a "false flag" attack that focused attention in the direction of Israel and the United States, while its actual purpose was different and broader. Other reports provide anecdotal evidence that point to the intelligence agencies of the two countries.
30. Most notably Ronen Bergman.
31. Taheri, *Persian Night*, 54–67.
32. In a contrasting view, George Friedman of Stratfor Global Intelligence argues that an effective popular uprising is unlikely to develop in the near future and that internal tensions within the government of Iran are not sufficient to create radical change. Friedman argues that the U.S. government's "fixation on internal evolutions in Iran has paralyzed American strategic thinking." *Agenda: With George Friedman on Iran*, 8 July 2011, www.stratfor.com (accessed 1 October 2012).
33. Gunzinger, "Outside-In," 16.
34. Ibid.
35. Ibid.
36. A conclusion implied in Majd's writings.
37. Gunzinger, "Outside-In,"19.
38. CNN, "Iranian Warships Sail into Mediterranean," 18 February 2012, http://www.cnn.com/2012/02/18/world/meast/iran-warships/index.html (accessed 1 October 2012)
39. U.S. Department of Defense, *Military Power of Iran*, 4.
40. Ibid., 22.

Chapter 8. Northeast Asia

1. "North Korea is, above all else, the world's most intense cult society with all the trappings of a nation-state." Robert L. Worden, ed., *North Korea: A Country Study*, 5th ed. (Washington, DC: Federal Research Division, Library of Congress, 2008), 77.
2. North Korea has created and imposed its own variant of Marxist-Leninist ideology called *juche*, which translates as "self-reliance" but in fact justifies isolation and loyalty to the supreme leader—Kim Il-sung and his successors. A brief summary of the effort to destroy religion and replace it with the cult of Kim Il-sung can be found in Victor Cha, *The Impossible State: North Korea, Past and Future* (New York: Ecco/HarperCollins, 2012), 72–74.

3. International Institute for Strategic Studies, "The Conventional Military Balance on the Korean Peninsula," extract from *North Korea's Weapons Programmes: A Net Assessment,* 2011, http://www.iss.org/publications/strategic-dossiers/north-korean-dossier/not-koreas-weapons-programes-a-net-assess/the-conventional-military-balance-on the kore/, 1 (accessed 1 September 2012).
4. Ibid., 219–20.
5. CBS News, "North Korea Successfully Launches Long-Range Rocket," 11 December 2012, http://www.cbsnews.com/8301–202_162–57558644/north-korea-successfully-launches-long-range-rocket/ (accessed 12 December 2012); Jack Kim and Mayumi Negishi, "North Korea Rocket Launch Raises Nuclear Stakes," Reuters, 12 December 2012, http://www.reuters.com/article/2012/12/12/us-korea-north-rocket-idUSBRE8BB02K20121212 (accessed 12 December 2012).
6. International Institute for Strategic Studies, "Conventional Military Balance," 6.
7. Sean O'Connor, "The North Korean SAM Network," *IMINT & Analysis* (blog), 12 June 2010 (updated 10 October 2012), http://geimt.blogspot.com/2010/06/north-korean-sam-network.html (accessed 1 December 2012).
8. International Institute for Strategic Studies, "Conventional Military Balance," 1.
9. Cha, *Impossible State*, 188.
10. Ibid., 170–71.
11. Ibid., 166.
12. This has been the official view of the South Korean government. Ibid., 403–12.
13. McKenzie provides a short vignette that captures the essence of this attack without ever actually mentioning Korea. See McKenzie, *Revenge of the Melians*, 77–78.
14. "Ten years ago these positions made a lot of sense; underground facilities proved very difficult to destroy even with precision guided munitions (PGM)—as demonstrated in GW1 and the Balkans. But we now live in the age of the "bunker buster." Planeman [pseud.], "Fortress North Korea," *Bluffer's Guide* (blog), February 9, 2008, http://www.militaryphotos.net/forums/showthread.php?128528-Bluffer-s-guide-Fortress-North-Korea, 4 (accessed 1 December 2012).
15. Cha, *Impossible State*, 429–30.
16. Worden, *North Korea*, 258.
17. Ibid., 260.

Chapter 9. Central Eurasia

1. In fairness it must be stated that Chechen rebels had committed deadly acts of terrorism in Moscow and other Russian Federation cities, acts to which any sovereign nation would respond as part of law enforcement. This, in fact, was the basis for a short-lived period of support by Putin for U.S. counterterrorism efforts under President George W. Bush. Additionally, it appears unclear whether the Chechen separatists, including Islamists and forces under the leadership of a former head of the Soviet air force, represented the will of the majority of Chechens—at least prior to the destruction of the resulting wars.
2. "Text of Newly-Approved Russian Military Doctrine: Report by Russian Presidential Website on 5 February" (not an official translation), Military Education Research Library Network, 5 February 2010, http://merln.ndu.edu/whitepapers/Russia2010_English.pdf, 3 (accessed 1 September 2012).
3. Lilia Shevtsova, *Lonely Power: Why Russia Has Failed to Become the West and the West Is Weary of Russia* (Washington, DC: Carnegie Endowment for International Peace: 2010), 145.
4. Eastern European members of NATO are convinced that there is a stockpile of tactical nuclear warheads in Kaliningrad, possibly for the Russian SS-23 ballistic missiles that have been moved there. Arms control advocates are skeptical. See Jorge Benitez, "NATO Allies Concerned about Nuclear Weapons in Kaliningrad," *NATO Source Alliance News Blog*, Atlantic Council, 11 February 2011, http://www.acus.org/natosource/nato-allies-concerned-about-nuclear-weapons-kaliningrad (accessed 1 September 2012); and Nikolai Sokov, "A Second Sighting of Russian Tactical Nukes in Kaliningrad, *CNS*, James Martin Center for Nonproliferation Studies, 15 February 2011, http://cns.miis.edu/stories/110215_kaliningrad_tnw.htm (accessed 1 September 2012).
5. Shevtsova, *Lonely Power*, 152.
6. Ibid.
7. John Drennan, "Russia Strengthens Its Hand in Central Asia," *IISS Voices: New Thinking from the International Institute for Strategic Studies* (blog), December 20, 2012, http://iissvoicesblog.wordpress.com/2012/12/20/russia-strengthens-its-hand-in-central-asia/ (accessed 21 December 2012).
8. Ibid. The conclusion that it will be tied in to the Russian system is my own.
9. Ibid.
10. Daniel Vajdic, "Russia's 'Shrewd' Central Asia Play," *The Diplomat*, 17 July 2012, http://thediplomat.com/flashpoints-blog/2012/07/17/russias-shrewd-central-asia-play (accessed 1 September 2012).
11. Ibid.

12. Marc Champion, "Putin's Clever Plan to Keep NATO in Afghanistan," *Bloomberg.com*, August 2, 2012, http://www.bloomberg.com/news/2012–08–02/putin-s-clever-plan-to-keep-nato-in-afghanistan.html (accessed 1 September 2012).
13. A brief summary on the SCO is Andrew Scheineson, "The Shanghai Cooperation Organization," Council of Foreign Relations, 24 March 2009, http://www.cfr.org/international-peace-and-security/shanghai-cooperation-organization/p10883/ (accessed 1 November 2012).
14. Andre C. Kuchins, "U.S.-Russia Relations: Constraints of Mismatched Strategic Outlooks," in *Russia after the Global Economic Crisis*, ed. Anders Aslund, Sergei Guriev, and Andrew C. Kuchins (Washington, DC: Peterson Institute for International Economics/Center for Strategic and International Studies/New Economic School, 2010), 251.
15. Shevtsova, *Lonely Power*, 145.
16. Ethnic cleansing appears to have occurred in Abkhazia. See WRITENET, *The Dynamics and Challenges of Ethnic Cleansing: The Georgia-Abkhazia Case, 1 August 1997*, Refworld, United Nations High Commissioner for Refugees, http://www.unhcr.org/refworld/docid/3ae6a6c54.html (accessed 1 November 2012).
17. Shevtsova, *Putin's Russia* (Washington, DC: Carnegie Endowment for International Peace, 2003), 234.
18. Robert Bridge, "Russia Tightens Border with Next-Gen Air Defense System," *Russia Times*, 15 October 2012, http://rt.com/politics/russia-air-defense-modernization-483/ (accessed 1 November 2012).

Chapter 10. Breaking Great Walls

1. David Hackett Fischer coined the term "the historian's fallacy" to describe the assumption by historians or readers of history "that a man who has a given historical experience knows it, when he has it, to be all that a historian would know it to be, with the advantage of historical perspective." See the discussion in David Aaronovitch, *Voodoo Histories: The Role of the Conspiracy Theory in Shaping Modern History* (New York: Riverhead, 2010), 268. This concept can be applied to analyses of revolutions in military affairs that paint advocates as champions and others as obstructionists. No one knows in advance if a revolution will actually succeed.
2. Roxana Tiron, "U.S. Places $42 Billion Bet on Carriers in China's Sights," *Bloomberg News*, 19 June 2012, www.bgov.com.
3. For a view that deterrence is situation-dependent, see Charles Krauthammer, "The 'Deterrence Works' Fantasy," *Washington Post*, 31 August 2012.
4. Sydney J. Freedberg, "Army Scrambles to Play Catch-Up on AirSea Battle," *AOL Defense* (blog), 7 June 2012, http://defense.aol.com/2012/06/07/

army-scrambles-to-play-catch-up-on-airsea-battle-invade-or-part/ (accessed 1 July 2012).

5. For a discussion of the sea-basing concept, see Sam J. Tangredi, "Sea Basing: Concept, Issues and Recommendations," *Naval War College Review* 64, no. 4 (Autumn 2011), 28–41.
6. An authoritative reference is Amy F. Wolf, *Conventional Prompt Global Strike and Long-Range Ballistic Missiles: Background and Issues*, Report for Congress R41464, (Washington, DC: Congressional Research Service, 2012).
7. David Axe, "Pentagon's Global Strike Weapon Stuck in Limbo: Congress Fears Accidental Nuclear War," *AOL Defense* (blog), 17 December 2012, http://defense.aol.com/2012/12/17/pentagons-global-strike-weapon-stuck-in-limbo-congress-fears-accidental-nuclear-war (accessed 20 December 2012); Matthew Fargo, "The Future of Conventional Prompt Global Strike," Center for Strategic and International Studies, 11 September 2012, https://csis.org/blog/future-conventional-prompt-global-strike (accessed 1 December 2012).

Bibliography

Aaronovitch, David. *Voodoo Histories: The Role of the Conspiracy Theory in Shaping Modern History*. New York: Riverhead, 2010.

Ahern, Dave. "AEI Sees Pentagon Afraid to Confront Growing Chinese Threat." *Defense Today*, 3 August 2005.

Ambrose, Stephen E. *Americans at War*. Jackson: University Press of Mississippi, 1997.

———. "Eisenhower, the Intelligence Community, and the D-Day Invasion." *Wisconsin Magazine of History* (Summer 1981): 261–77.

Atkinson, Rick. *Crusade: The Untold Story of the Persian Gulf War*. Boston: Houghton Mifflin, 1993.

Axe, David. "Pentagon's Global Strike Weapon Stuck in Limbo: Congress Fears Accidental Nuclear War." *AOL Defense* (blog), 17 December 2012. http://defense.aol.com/2012/12/17/pentagons-global-strike-weapon-stuck-in-limbo-conress-fears-accidental-nuclear-war.

Barber, Arthur H., III, and Delwyn L. Gilmore. "Maritime Access: Do Defenders Hold All the Cards?" *Defense Horizons*, no. 4 (October 2001). http://www.ndu.edu/CTNSP/docUploaded/DH_04.pdf.

Barnett, Roger W. *Asymmetrical Warfare: Today's Challenges to U.S. Military Power*. Washington, DC: Brassey's, 2002.

Barnett, Thomas P. M. *Great Powers: America and the World after Bush*. New York: G. P. Putnam's Sons, 2009.

———. *The Pentagon's New Map: War and Peace in the Twenty-First Century*. New York: Berkley, 2004.

Beaver, Paul. "China Develops Anti-Satellite Laser System." *Jane's Defence Weekly*, 2 December 1998, 18.

Beck, Daniel C., Yash Holbrook, Randall Greenwalt, David R. Beachley, and John A. Battilega. *Global Perspectives on the Revolution in Military Affairs: Selected Asymmetric Responses; China, France, Russia, Iran, India*. Colorado Springs, CO: SAIC Foreign Systems Research Center, 1998.

Benitez, Jorge. "NATO Allies Concerned about Nuclear Weapons in Kalinigrad." *NATO Source Alliance News Blog* (Atlantic Council), 11 February 2011. http://www.acus.org/natosource/nato-allies-concerned-about-nuclear weapons-kalinigrad.

Bennett, John T. "Chinese Buildup of Cyber, Space Tools Worries U.S." *Defense News*, 13 January 2010. http://www.defensenews.com/article/20100113/DEFSECT04/1130301/Chinese-Buildup-Cyber-Space-Tools-Worries-US-.

Bergman, Ronen. *The Secret War with Iran: The 30-Year Clandestine Struggle against the World's Most Dangerous Terrorist Power*. New York: Free Press, 2008.

Blumenthal, Daniel. "China's Grand Strategy." *Foreign Policy: Shadow Government* (blog), 29 April 2010. http://www.foreignpolicy.com/posts/2010/04/29/china_s_grand_strategy.

Bodansky, Yossef. "The DPRK-RoK Crisis: Pyongyang, Once Again, Took Decisive, Successful Steps to Win Attention and Respect." *Defense and Foreign Affairs Special Analysis* 27, no. 75 (25 November 2010).

Boling, James L. "Rapid Decisive Operations: The Emperor's New Clothes in Modern Warfare." In *Essays 2002: Chairman of the Joint Chiefs of Staff Essay Competition*. Washington, DC: National Defense University Press, 2002, 41–62.

Bosone, John. *Airpower and the Establishment of Sea Control in an Antiaccess Environment*. Newport, RI: Naval War College, 2008.

Bowdish, Randall G., and Bruce Woodyard. "A Naval Concepts-Based Vision for Space," U.S. Naval Institute *Proceedings* 125, no. 1 (January 1999): 50–53.

Bowie, Chistopher J. *The Anti-Access Threat to Theater Air Bases*. Washington, DC: Center for Strategic and Budgetary Assessments, 2002.

Bracken, Paul. *Fire in the East: The Rise of Asian Military Power and the Second Nuclear Age*. New York: HarperCollins, 1999.

Bradsher, Keith. "China Drawing High-Tech Research from U.S." *New York Times*, 17 March2010. http://www.nytimes.com/2010/03/18/business/global/18research.html?th&emc+th.

———. "China Uses Rules on Global Trade to Its Advantage." *New York Times*, 14 March 2010. http://www.nytimes.com/2010/03/15/business/global/15yuan.html?th&emc+th.

Bridge, Robert. "Russia Tightens Border with Next-Gen Air Defense System." *Russia Times*, 15 October 2012. http://rt.com/politics/russia-air-defense-modernization-483/.

Bruger, Steven J., "Not Ready for the First Space War: What about the Second?" *Naval War College Review* 48, no. 1 (Winter 1995): 73–83.

Brzezinski, Zbigniew. "Living with China." *The National Interest*, no. 59 (Spring 2000): 5–21.

Bungay, Stephen. *The Most Dangerous Enemy: A History of the Battle of Britain*. London: Aurum, 2000.

Burns, John F. "Confrontation in the Gulf: Behind Iraq's Resolve, a Militia of the Irregular." *New York Times*, 3 October 1990. www.nytimes.com/1990/10/30/world/confrontation-in-the-gulf-behind-iraq-s-resolve-a-militia-of-the-irregular.html.

———. "India's Defense Minister Calls U.S. Defense Policies 'Hypocritical.'" *New York Times*, 18 June 1998.

Bush, Richard C., and Michael E. O'Hanlon. *A War like No Other: The Truth about China's Challenge to America*. Hoboken, NJ: John Wiley, 2007.

Bussert, James C., and Bruce A. Elleman. *People's Liberation Army Navy: Combat Systems Technology, 1949–2010*. Annapolis, MD: Naval Institute Press, 2011.

Cagle, Malcolm W., and Frank A. Manson. *The Sea War in Korea*. Annapolis, MD: Naval Institute Press, 1957.

Callwell, C. E. *Military Operations and Maritime Preponderance: Their Relations and Interdependence*. Annapolis, MD: Naval Institute Press, 1996 (orig. 1905).

Cappacio, Tony. "China's New Missile May Create a 'No-Go Zone' for U.S. Fleet." Bloomberg.com, 17 November 2009.

Carman, Jesse C. "Economic and Strategic Implications of Ice-Free Arctic Seas." In *Globalization and Maritime Power*, edited by Sam J. Tangredi, 171–87. Washington, DC: National Defense University Press, 2002.

CBS News. "North Korea Successfully Launches Long-Range Rocket." 11 December 2012. http://www.cbsnews.com/8301–202_162–57558644/north-korea-successfully-launches-long-range-rocket/.

Cha, Victor. *The Impossible State: North Korea, Past and Future*. New York: Ecco/HarperCollins, 2012.

Champion, Marc. "Putin's Clever Plan to Keep NATO in Afghanistan." Bloomberg.com, 2 August 2012. http://www.bloomberg.com/news/2012–08–02/putin-s-clever-plan-to-keep-nato-in-afghanistan.html.

Chant, Chistopher. *The Military History of the United States: Revolutionary and Early American Wars*. New York: Webster's Home Library, 1997.

China Defense Blog. "First Photos of S-400 in China." http//china-defense.blogspot.com/2009/05/first-photos-of-s-400-in-china.html.

Chivers, C. J. "Work as Usual for U.S. Warship after Warning by Iran." *New York Times*, 4 January 2012. http://www.nytimes.com/2012/01/05/

world/middleeast/work-as-usual-for-uss-john-c-stennis-after-warning-by-iran/.

Churchill, Winston. *The Second World War, Volume 2: Their Finest Hour.* London: Cassell, 1949.

Clapp, Michael, and Ewen Southby-Tailyour. *Amphibious Assault Falklands: The Battle of San Carlos Water.* Barnsley, UK: Pen & Sword, 2007.

Cliff, Roger, Mark Burles, Michael S. Chase, Derek Eaton, and Kevin L. Pollpeter. *Entering the Dragon's Lair: Chinese Antiaccess Strategies and Their Implications for the United States.* Santa Monica, CA: RAND Corp., 2007.

CNN. "Iranian Warships Sail into Mediterranean," 18 February 2012. http://www.cnn.com/2012/02/18/world/meast/iran-warships/index.html.

Cohen, Eliot A., and John Gooch. *Military Misfortunes: The Anatomy of Failure in War.* New York: Free Press, 2005.

Colby, Elbridge. "The Role of Nuclear Weapons in the Global Security Environment." *Global Trends 2030* (blog), June 29, 2012. http://gt20230.com/2012/06/29/the-role-of-nuclear-weapons-in-the- future security-environment-2/.

Cole, Bernard D. *The Great Wall at Sea, Second Edition: China's Navy in the Twenty-First Century.* Annapolis, MD: Naval Institute Press, 2010.

Collins, Gabriel B. "China's Dependence on the Global Maritime Commons." In *China, the United States and 21st-Century Sea Power: Defining a Maritime Security Partnership*, edited by Andrew S. Erickson, Lyle Goldstein, and Nan Li, 14–37. Annapolis, MD: Naval Institute Press, 2010.

Collins, John M. *Grand Strategy: Principles and Practices.* Annapolis, MD: Naval Institute Press, 1973.

Corbett, Julian. *History of the Great War: Naval Operations*, vols. 1 and 2. London: Longmans, Green, 1920, 1921.

Cote, Owen R., Jr. *Assuring Access and Projecting Power.* Boston: MIT Security Studies Program, 2001.

Couhat, Jean Labaye, and A. D. Baker, eds. *Combat Fleets of the World 1986/87.* Annapolis, MD: Naval Institute Press, 1986.

Craig, William. *The Fall of Japan.* New York: Dial, 1967.

Cronin, Patrick, and Paul Giarra, "China's Dangerous Arrogance: An Increasingly Assertive China Is Creating Its Own Monroe Doctrine for Asia's Seas—and Threatening Longstanding Freedoms." *The Diplomat*, July 23, 2010. http://thediplomat.com/flashpoints-blog/2010/07/23/china%e2%80%99s-dangerous-arrogance/.

Danzig, Richard. *The Big Three: Our Greatest Security Risks and How to Address Them.* New York: Center for International Political Economy, 1999.

Davis, Paul K., Jimmie McEver, and Barry Wilson. *Measuring Interdiction Capabilities in the Presence of Anti-Access Strategies: Exploratory Analysis to Inform Adaptive Strategy for the Persian Gulf.* Santa Monica, CA: RAND Corp., 2002.

Defense Science Board. *Report of the Defense Science Board Task Force on Strategic Mobility.* Washington, DC: Office of the Secretary of Defense, 1996.

Deighton, Len. *Battle of Britain.* London: George Rainbird, 1980.

———. *Fighter: The True Story of the Battle of Britain.* New York: Knopf, 1978.

Denmark, Abraham M., and James Mulvenon. *Contested Commons: The Future of American Power in a Multipolar World.* Washington, DC: Center for a New American Security, 2010.

de Souza, Phillip. *The Greek and Persian Wars 499–386 BC.* Oxford: Osprey, 2003.

Dossel, Will. "Ballistic Missile Defense." In *Securing Freedom in the Global Commons*, edited by Scott Jasper, 115–30. Stanford, CA: Stanford University Press, 2010.

Drennan, John. "Russia Strengthens Its Hand in Central Asia." *IISS Voices: New Thinking from the International Institute for Strategic Studies* (blog), 20 December 2012. http://iissvoicesblog.wordpress .com/2012/12/20/russia-strengthens-its-hand-in-central-asia/.

Economic Collapse Blog. "11 Reasons Why North Korea Is the Most Bizarre Nation on Earth." 24 November 2010. http://theeconomiccollapseblog/ archives/11-reasons-why-north-korea-is-the-most-bizarre-nation-on-earth.

Economist, The. "Frying Tonight: Warfare Is Changing as Weapons That Destroy Electronics, Not People, Are Deployed on the Field of Battle." 15 October 2011. http://www.economist.com/node/21532245.

Erickson, Andrew S. "Are China's Near Seas 'Anti-Navy' Capabilities Aimed Directly at the United States?" *Information Dissemination* (blog), 14 June 2012. www.informationdissemination.net/2012/06/are-chinas-near-seas-anti-navy.htm.

Erickson, Andrew S., Lyle J. Goldstein, and Nan Li, eds. *China, the United States and 21st Century Sea Power: Defining a Maritime Security Partnership.* Annapolis, MD: Naval Institute Press, 2010.

Erickson, Andrew S., Lyle J. Goldstein, and Carnes Lord, eds. *China Goes to Sea: Maritime Transformation in Comparative Historical Perspective.* Annapolis, MD: Naval Institute Press, 2009.

Erickson, Andrew S., and David D. Yang. "On the Verge of a Game Changer." U.S. Naval Institute *Proceedings* 135, no. 5 (May 2009): 26–32.

———. "Using the Land to Control the Seas: Chinese Analysts Consider the Antiship Ballistic Missile." *Naval War College Review* 62, no. 4 (Autumn 2009): 53–86.

Ewing, Phillip. "The Rise and Fall of Air-Sea Battle." *DoD Buzz: Online Defense and Acquisition Journal*, 17 May 2012. www.dodbuzz.com/2012/05/17/the-rise-and-fall-of-air-sea-battle/.

"Excerpts from Iraqi Document on Meeting with U.S. Envoy." *New York Times International*, 23 September 1990. http://msuweb.montclair.edu/~furrg/glaspie.html.

Fargo, Matthew. "The Future of Conventional Prompt Global Strike." Center for Strategic and International Studies, 11 September 2012. https://csis.org/blog/future-conventional-prompt-global-strike.

Feaver, Peter. "Should the Obama Administration Release Another National Security Strategy This Term?" *Shadow Foreign Policy* (blog), 3 February 2012. http://shadow.foreignpolicy.com/posts/2012/02/03/should_the_obama_administration_relase_another_national_security_strategy_this_ter.

Finlan, Alistair. *The Gulf War 1991*. Oxford: Osprey, 2003.

Fisher, Richard D., Jr. "China's Scary Space Ambitions." *Wall Street Journal*, 20 January 2010. http://online.wsj.com/article/SB10014240527487043201045750143414636158 67.html.

Fleming, Peter. *Operation Sea Lion*. New York: Simon & Schuster, 1957.

Floumoy, Michele, and Shawn Brimley. "The Contested Commons." U.S. Naval Institute *Proceedings* 135, no. 7 (July 2009): 16–21.

Ford, Douglas. *The Elusive Enemy: U.S. Naval Intelligence and the Imperial Japanese Fleet*. Annapolis, MD: Naval Institute Press, 2011.

Freedberg, Sidney J. "Army Scrambles to Play Catch-Up on AirSea Battle." *AOL Defense*, 7 June 2012. http://defense.aol.com/2012/06/07/army-scrambles-to-play-catch-up-on-airsea-battle-invade-or-part/.

———. "Cartwright Targets F-35, AirSea Battle: Warns of $250B More Cuts." *AOL Defense*, 15 May 2012. http://defense.aol.com/2012/05/15/cartwright-savages-f-35-airsea-battle-warns-of-250-billion-mo/.

———. "Iran Mine Threat Scares Navy: CNO Scrambles to Fix Decades of Neglect." *AOL Defense*, 4 May 2012. http://defense.aol.com/2012/05/04/iran-mine-threat-scares-navy-cno-scrambles-to-fix-decades-of-neglect.html.

Freedman, Lawrence, and Efraim Karsh. *The Gulf Conflict 1990–1991: Diplomacy and War in the New World Order*. Princeton, NJ: Princeton University Press, 1993.

Friedman, George. "Agenda: With George Friedman on Iran." *Stratfor*, 8 July 2011. www.stratfor.com.

———. *America's Secret War: Inside the Hidden Worldwide Struggle between America and Its Enemies*. New York: Doubleday, 2004.

———. "Poland's Strategy." *Stratfor*, 28 August 2012. www.stratfor.com.

Friedman, George, and Meredith Friedman. *The Future of War.* New York: St. Martin's Griffin, 1996.

Friedman, Norman. "The Falklands War: Lessons Learned and Mislearned."*Orbis* 26, no. 4 (Winter 1983): 907–40.

———. "Globalization of Antiaccess Strategies." In *Globalization and Maritime Power*, edited by Sam J. Tangredi, 487–502. Washington, DC: National Defense University Press, 2002.

———. *The U.S. Maritime Strategy.* Annapolis, MD: Naval Institute Press, 1988.

Gatchel, Theodore L. *At the Water's Edge: Defending against the Modern Amphibious Assault.* Annapolis, MD: Naval Institute Press, 1996.

Gerth, Jeff. "Administration Rethinking $650 Million China Satellite Deal. *New York Times*, 18 June 1998.

Gilbert, Martin. *Winston S. Churchill, Volume 6: Finest Hour, 1939–1941.* London: Heinemann, 1983.

Global Security.org. "China Building EMP Arms in Case of Conflict over Taiwan: Report." *Global Security Report*, July 24, 2011. http://www.globalsecurity.org/wmd/library/news/china/2011/china-110724-cna01.htm.

Goodwin, Paul H. B. "PLA Doctrine and Strategy: Mutual Apprehension in Sino-American Military Planning." In *The People's Liberation Army and China in Transition*, edited by Stephen J. Flanagan and Michael E. Marti, 261–84. Washington, DC: National Defense University Press, 2003.

Gorshkov, S. G. *The Sea Power of the State.* Annapolis, MD: Naval Institute Press, 1979.

Gray, Colin S. *National Security Dilemmas: Challenges and Opportunities.* Washington, DC: Potomac Books, Inc., 2009.

Green, Michael. *America's Grand Strategy in Asia: What Would Mahan Do?* Sydney: Lowy Institute McArthur Asia Security Project, 2010.

Greenfield, Kent Roberts. *American Strategy in World War II: A Reconsideration.* Baltimore: Johns Hopkins University Press, 1963.

Grevatt, Jon. *Jane's Missiles and Rockets.* London: Jane's Defence Group, 2007.

Grove, Eric. *The Future of Sea Power.* Annapolis, MD: Naval Institute Press, 1990.

Gruselle, Bruno. *Cruise Missiles and Anti-Access Strategies.* Paris: Foundation pour la Recherche Stratégique, 2006. http://www.frstrategie.org/barreFRS/publications/rd/RD_20060601_eng.pdf

Gunzinger, Mark. "Outside-In: Defeating Iran's Anti-Access and Area-Denial Threat" (briefing slides). Washington, DC: CSBA, 17 January 2012.

Gunzinger, Mark, with Chris Dougherty. *Outside-In: Operating from Range to Defeat Iran's Anti-Access and Area Denial Threats.* Washington, DC: CSBA, 2011.

Halpin, Tony. "Russia Warns of War within a Decade over Arctic Oil and Gas Riches." *The Times*, 14 May 2009. http://www.timesonline.co.uk/tol/news/environment/article6283130.ece.

Halsey, Francis Whiting. *The Literary Digest History of the World War*, vol. 10. New York: Funk & Wagnalls, 1920.

Handel, Michael. *Weak States in the International System.* London: Frank Cass, 1981.

Hart, Rodger. *Battle of the Spanish Armada.* East Sussex, UK: Wayland, 1973.

Hawkins, William R. "Desperate to Trade with the Enemy?" *Family Security Matters* (blog), 3 May 2010. http://familysecuritymatters.org/publications/id.6120/pub_detail.asp.

Helprin, Mark. "Farewell to America's China Station." *Wall Street Journal*, 17 May 2010.

Henry, Ryan. "Transforming the U.S. Global Defense Posture." *Naval War College Review* 59, no. 2 (Spring 2006): 13–28.

Her Majesty's Government, Ministry of Defence, Development, Concepts, and Doctrine Center. *The DCDC Strategic Trends Programme 2007–2036*. London: Ministry of Defence, 2007.

Herodotus. *The Histories.* Translated by George Rawlinson. New York: Knopf, 1997.

Hime, Douglas M. *The 1982 Falklands-Malvinas Case Study*, NWC 1036. Newport, RI: U.S. Naval War College, 2010.

Hoffman, Frank G. "Hybrid Threats: Neither Omnipotent nor Unbeatable." *Orbis* 54, no. 3 (Summer 2010): 446-452

Holland, James. *The Battle of Britain: Five Months That Changed History; May–October 1940.* New York: St. Martin's Griffin, 2012.

Hooper, Craig, and Christopher Albon. "Get off the Fainting Couch." U.S. Naval Institute *Proceedings* 136, no. 4 (April 2010): 42–46.

Hoyt, Edwin P. *Yamamoto: The Man Who Planned Pearl Harbor.* New York: McGraw-Hill, 1990.

Hresko, Eric. *Effects-Based Operations: A Valid Concept for Operations in an Anti-Access Environment* (thesis). Newport, RI: Naval War College, 2008.

Hughes, Wayne P., Jr. *Fleet Tactics and Coastal Combat*, 2nd ed. Annapolis, MD: Naval Institute Press, 2000.

Hybel, Alex Roberto. *The Logic of Surprise in International Conflict.* Lexington, MA: Lexington Books, 1986.

Hyten, John E. *A Sea of Peace or a Theater of War: Dealing with the Inevitable Conflict in Space.* ACDIS Occasional Paper. Champaign, IL: University of Illinois at Urbana-Champaign, 2000.

IHS Jane's, "First Photos of S-400 in China." In *Jane's Strategic Weapon Systems* (London: Jane's Defence Group, 2008.

———. "Samsung-Thales Begins M-SAM Radar Production." In *Jane's Missiles and Rockets.* London: Jane's Defence Group, 2007),

International Institute for Strategic Studies. "The Conventional Military Balance on the Korean Peninsula." Extract from *North Korea's Weapons Programmes: A Net Assessment.* London: International Institute for Strategic Studies, 2011. http://www.iss.org/publications/strategic-dossiers/north-korean-dossier/not-koreas-weapons-programes-a-net-assess/the-conventional-military-balance-on the kore/.

International Institute for Strategic Studies, *The Military Balance 2012.* London: Routledge, 2012.

Irving, David. *Hitler's War.* New York: Viking, 1977.

Jackson, Paul, ed. *Jane's All the World's Aircraft: 2012–2013.* London: IHS Jane's Information Group, 2012.

Jacobs, Andrew. "In Leaked Lecture, Details of China's News Cleanups." *New York Times,* 3 June 2010. http://www.nytimes.com/2010/06/04/world/asia/04china.html?th&emc+th.

Jafarzadeh, Alireza. *The Iran Threat: President Ahmadinejad and the Coming Nuclear Crisis.* New York: Palgrave Macmillan, 2007.

Jasper, Scott, ed. *Conflict and Cooperation in the Global Commons.* Washington, DC: Georgetown University Press, 2012.

Joseph, Robert G., and John F. Reichart. *Deterrence and Defense in a Nuclear, Biological, and Chemical Environment.* Washington, DC: National Defense University, Institute for National Strategic Studies, 1995.

Jumper, John P. "Expeditionary Air Force: A New Culture for a New Century." Air Force Association, 26 February 1998. www.afa.org/afe/pub/ol19.asp.

Kaplan, Robert. "The Geography of Chinese Power." *Foreign Affairs,* May/June 2010: 22–41.

———. *The Revenge of Geography.* New York: Random House, 2012.

Katzman, Kenneth, Neelesh Nerurkar, Ronald O'Rourke, R. Chuck Mason, and Michael Ratner. *Iran's Threat to the Strait of Hormuz,* Report for Congress R42335. Washington, DC: Congressional Research Service, 2012.

Kemp, Peter. *The Campaign of the Spanish Armada.* New York: Facts on File, 1988.

Kessler, Glenn. "Ex-Envoy Details Hussein Meeting." *Washington Post,* 3 April 2008. http://www.washingtonpost.com/wp-dyn/content/article/2008/04/02/AR2008040203485.html.

Keyes, Roger. *The Fight for Gallipoli*. London: Eyre & Spottiswoode, 1934.

Kieser, Egbert. *Hitler on the Doorstep: Operation "Sea Lion"; The German Plan to Invade Britain, 1940*. Annapolis, MD: Naval Institute Press, 1997.

Kim, Jack, and Mayumi Negishi. "North Korea Rocket Launch Raises Nuclear Stakes." Reuters, 12 December 2012. http://www.reuters.com/article/2012/12/12/us-korea-north-rocket-idUSBRE8BB02K20121212.

Knox, Dudley W. *The Naval Genius of George Washington*. Boston: Houghton Mifflin, 1932.

Korda, Michael. *With Wings like Eagles: A History of the Battle of Britain*. New York: HarperCollins, 2009.

Krause, Clifford. "Oil Price Would Skyrocket if Iran Closed the Strait of Hormuz." *New York Times*, 4 January 2012. http://www.nytimes.com/2012/01/05/business/oil-price-would-skyrocket-if-iran-closed-the-strait-of-hormuz/.

Krauthammer, Charles. "The 'Deterrence Works' Fantasy." *Washington Post*, 31 August 2012.

Krepinevich, Andrew F., Jr. "Cavalry to Computer: The Pattern of Military Revolutions." *National Interest* 37 (Fall 1994): 30–42.

———. *The Military-Technical Revolution: A Preliminary Assessment*. Washington, DC: Center for Strategic and Budgetary Assessments, 2002.

———. *7 Deadly Scenarios: A Military Futurist Explores War in the 21st Century*. New York: Bantam, 2009.

———. *Why AirSea Battle?* Washington, DC: Center for Strategic and Budgetary Assessments, 2010.

Krepinevich, Andrew, Barry Watts, and Robert Work. *Meeting the Anti-Access and Area Denial Challenge*. Washington, DC: Center for Strategic and Budgetary Assessments, 2003.

Kuchins, Andrew C. "U.S.-Russia Relations: Constraints of Mismatched Strategic Outlooks." In *Russia after the Global Economic Crisis*, edited by Anders Aslund, Sergei Guriev, and Andrew C. Kuchins, 241–56. Washington, DC: Peterson Institute for International Economics/Center for Strategic and International Studies/New Economic School, 2010.

LaFraniere, Sharon, and Jonathan Ansfield. "China Alarmed by Security Threat from Internet." *New York Times*, 11 February 2010. http://www.nytimes.com/2010/02/12/world/asia/12cyberchina.html?ref=world.

LaGrone, Sam. "Pentagon's 'Air-Sea Battle' Plan Explained. Finally." *Wired.com Danger Room*, 6 August 2012. www.wired.com/dangerroom/2012/08/air-sea-battle-2/.

Larson, Eric V., Derek Eaton, Paul Elrick, Theodore Karasik, Robert Klein, Sherrill Lingel, Brian Nichiporuk, Robert Uy, and John Zavadil. *Assuring Access in Key Strategic Regions: Toward a Long-Term Strategy*. Santa Monica, CA: RAND Corp., 2004.

Lasater, Martin L. "Conflict in the Taiwan Strait: The American Response." *Taiwan Security Research*, February 2000.

Lewin, Ronald. *The American Magic: Codes, Ciphers and the Defeat of Japan*. New York: Farrar, Straus and Giroux, 1982.

Liddell Hart, Basil, and Adrian Liddell Hart. *The Sword and the Pen*. New York: Thomas Y. Crowell, 1976.

Looney, Robert E. "Market Effects of Naval Presence in a Globalized World: A Research Summary." In *Globalization and Maritime Power*, edited by Sam J. Tangredi, 103–31. Washington, DC: National Defense University Press, 2002.

Macintyre, Donald. *The Naval War against Hitler*. New York: Charles Scribner's Sons, 1971.

MacMillan, Margaret. *Nixon, Kissinger, and the Opening to China*. New York: Random House, 2008.

Mahan, Alfred Thayer Mahan. *Lessons of the War with Spain and Other Articles*. Boston: Little, Brown, 1899.

Mahnken, Thomas G. "China's Anti-Access Strategy in Historical and Theoretical Perspective." *Journal of Strategic Studies* 34, no. 3 (June 2011): 299–323.

———. "Deny U.S. Access?" U. S. Naval Institute *Proceedings* 124, no. 9 (September 1998): 36–39.

Majd, Hooman. *The Ayatollahs' Democracy: An Iranian Challenge*. New York: Norton, 2010.

Manke, Robert C., and Raymond J. Christian. *Asymmetry in Maritime Access and Undersea Anti-Access/Area-Denial Strategies*, Technical Report 11,826. Newport, RI: Naval Undersea Warfare Center, 2007.

Manor, Mike, and Kurt Neuman. "Space Assurance." In *Securing Freedom in the Global Commons*, edited by Scott Jasper, 99–114. Stanford, CA: Stanford University Press, 2010.

Markoff, John, and David Barboza. "Paper in China Sets Off Alarms in U.S." *New York Times*, 20 March 2010. www.nytimes.com/2010/03/21/world/asia/21grid.html.

Martin, Colin, and Geoffrey Parker. *The Spanish Armada*. New York: Norton, 1988.

Massie, Robert K. *Castles of Steel: Britain, Germany, and the Winning of the Great War*. New York: Random House, 2003.

Matray, James I. "Dean Acheson's Press Club Speech Reexamined." *Journal of Conflict Studies* 22, no. 1 (Spring 2002). http://journals.hil.unb.na/index.php/jcs/article/view366/578.

Matthews, Lloyd J., ed. *Challenging the United States Symmetrically and Asymmetrically: Can America Be Defeated?* Carlisle Barracks, PA: U.S. Army War College, 1998.

Mattingly, Garrett. *The Armada*. Boston: Houghton Mifflin, 1959.

Mattis, James N., *Statement of General James N. Mattis, USMC, Commander, United States Joint Forces Command, before the Senate Armed Services Committee*. Committee on Senate Armed Services, 111th Congress, sess. 2, 9 March 2010. http://www.jfcom.mil/newslink/storyarchive/2010/sp030910.html.

McIvor, Anthony D., ed. *Rethinking the Principles of War*. Annapolis, MD: Naval Institute Press, 2005.

McKee, Alexander. *From Merciless Invaders: An Eyewitness Account of the Spanish Armada*. New York: Norton, 1963.

McKenzie, Kenneth F., Jr. *The Revenge of the Melians: Asymmetric Threats and the Next QDR*, McNair Paper 62. Washington, DC: National Defense University Press, 2000.

Middlebrook, Martin. *The Fight for the Malvinas: The Argentine Forces in the Falklands War*. London: Viking, 1989.

———. *Operation Corporate: The Story of the Falklands War, 1982*. London: Viking, 1985.

Miller, Edward S. *War Plan Orange: The U.S. Strategy to Defeat Japan, 1897–1945*. Annapolis, MD: Naval Institute Press, 1991.

Minnick, Wendell. "China Builds First Anti-Ship Ballistic Missile Base?" *Defense News*, 6 August 2010.

Montagu, Ewen. *Beyond Top Secret Ultra*. New York: Coward, McCann & Geoghegan, 1978.

Moorehead, Alan. *Gallipoli*. Hertfordshire, UK: Wordsworth, 1997.

Morison, Samuel Eliot. *Strategy and Compromise*. Boston: Little, Brown, 1958.

———. *The Two-Ocean War: A Short History of the United States Navy in the Second World War*. Boston: Little, Brown, 1963.

Murdoch, Clark A. "The Navy in an Antiaccess World." In *Globalization and Maritime Power*, edited by Sam J. Tangredi, 473–86. Washington, DC: National Defense University Press, 2002.

Myers, Gene. "Getting to the Fight: Aerospace Forces and Anti-Access Strategies." *Air & Space Power Journal*, 27 March 2001. http://www.airpower.maxwell.af.mil/airchronicles/cc/myers01.html.

National Defense Panel. *Transforming Defense: National Security in the 21st Century*. Arlington, VA: National Defense Panel 1997.

Ochmanek, David. "The Air Force: The Next Round." In *Transforming America's Military*, edited by Hans Binnendijk, 159–92. Washington, DC: National Defense University Press, 2002.

O'Connell, Robert L. "A Useful Navy for 2017: What Can Naval History Tell Us?" In *New Interpretations in Naval History: Selected Papers from the Thirteenth Naval History Symposium*, edited by William M. McBride and Eric P. Reed, 308–21. Annapolis, MD: Naval Institute Press, 1998.

O'Connor, Sean. "The North Korean SAM Network." *IMINT & Analysis* (blog), 12 June 2010 (updated 10 October 2012). http://geimt.blogspot.com/2010/06/north-korean-sam-network.html.

O'Halloran, James C., ed. *Jane's Land-Based Air Defence*. London: IHS Jane's Information Group, 2011.

O'Rourke, Ronald. *China Naval Modernization: Implications for U.S. Navy Capabilities; Background and Issues for Congress*. Washington, DC: Congressional Research Service, 2010.

________. *Naval Modernization: Implications for U.S. Navy Capabilities; Background and Issues for Congress*. Washington, DC: Congressional Research Service, 2009.

Overy, Richard. *Battle of Britain: The Myth and Reality*. New York: Norton, 2001.

Pant, Harsh V. "Many in Denial over China's Quest for Bases." *Japan Times Online*, 12 February 2010. http://search.japantimes.co.jp/cgi-bin/eo20100212a1.html.

Parker, Frederick D. *A Priceless Advantage: U.S. Navy Communications Intelligence and the Battles of Coral Sea, Midway and the Aleutians*. Washington, DC: National Security Agency, 2001.

Pehrson, Christopher J. *String of Pearls: Meeting the Challenge of China's Rising Power across the Asian Littoral*. Carlisle Barracks, PA: Strategic Studies Institute, U.S. Army War College, 2006.

Perlez, Jan. "Singaporean Tells China U.S. Is Not in Decline." *New York Times*, 6 September 2012. www.nytimes.com/2012/09/07world/asia/singapores-prime-minister-warns-china-on-view-of-US.htm.

Perry, James D. "Operation Allied Force: The View from Beijing." *Aerospace Power Journal* 24, no. 3 (Summer 2000): 79–91.

Peters, Ralph. *Endless War: Middle-Eastern Islam vs. Western Civilization*. Mechanicsburg, PA: Stackpole, 2010.

———. *Fighting for the Future: Will America Triumph?* Mechanicsburg, PA: Stackpole, 1999.

Pillsbury, Michael, ed. *China Debates the Future Security Environment*. Washington, DC: National Defense University Press, 2000.

———. *Chinese Views of Future Warfare*, rev. ed. Washington, DC: National Defense University Press, 1998.

Planeman [pseud.]. "Fortress North Korea." *Bluffer's Guide* (blog), 9 February 2008. http://www.militaryphotos.net/forums/showthread.php?128528-Bluffer-s-guide-Fortress-North-Korea.

Plutarch. *The Lives of the Noble Grecians and Romans*. Translated by John Dryden. Revised by Arthur Hugh Clough. New York: Modern Library, orig. 1864.

Posen, Barry R. "Command of the Commons: The Military Foundation for U.S. Hegemony." *International Security* 28, no. 1 (Summer 2003): 5–46.

Potter, E. B., and Chester W. Nimitz. *Sea Power: A Naval History*. Englewood Cliffs, NJ: Prentice Hall, 1960.

Prange, Gordon. *At Dawn We Slept*. New York: Penguin, 1982.

Qiao Liang and Wang Xiangsui. *Unrestricted Warfare: China's Master Plan to Destroy America*. Los Angeles: Pan American, 2002.

Raleigh, Walter. *The Works of Walter Raleigh, Kt.*, vol. 8. Oxford, UK: Oxford University Press, 1829.

Ranft, Bryan, and Geoffrey Till. *The Sea in Soviet Strategy*. Annapolis, MD: Naval Institute Press, 1983.

Raska, Michael. "Air-Sea Battle Debate: Operational Consequences and Allied Concerns." *Defense News*, 30 October 2012. http://www.defensenews.com/article/20121030/DEFFEAT05/310300008/Air-Sea-Battle-Debate?odyssey=nav%7Chead.

Richardson, Lewis Fry. *Statistics of Deadly Quarrels*. Pittsburgh: Boxwood, 1960.

Robinson, Derek. *Invasion 1940: The Truth about the Battle of Britain and What Stopped Hitler*. New York: Da Capo, 2005.

Ruddeno, Victor. *Gallipoli: Attack from the Sea*. New Haven, CT: Yale University Press, 2008.

Saunders, Stephen, ed. *Jane's Fighting Ships: 2012–2013*. London: IHS Jane's Information Group, 2012.

Schaller, Erich Udo. *Naval Surface Force Protection in a Long War: A Consideration of the Anti-Access Threat* (thesis). Newport, RI: U.S. Naval War College, 2006.

Scheineson, Andrew. "The Shanghai Cooperation Organization." *Council on Foreign Relations* (blog), 24 March 2009. http://www.cfr.org/international-peace-and-security/shanghai-cooperation-organization/p10883/.

Schiffrin, Nick, and Matthew McGarry. "Iran, US Flex Military Muscles in Persian Gulf: US Carries Out Vast Allied Anti-Mine Exercise, Iran Test Fires Anti-Warship Missiles." ABC News, 25 September 2012. abcnews.go.com.

Schmitt, Eric. "Chinese Suddenly Improved Rocket Launches, Expert Says." *New York Times*, June 18, 1998.

Schwartz, Norton A., and Jonathan W. Greenert. "Air-Sea Battle: Promoting Stability in an Era of Uncertainty." *The American Interest*, 20 February 2012. www.the-american-interest.com/article.cfm?piece=1212.

Shanker, Thom, and Rick Gladsone. "Iran Fired on Military Drone in First Such Attack, U.S. Says." *New York Times*, 9 November 2012.

Shankland, Peter. *Dardanelles Patrol*. New York: Scribner, 1964.

Shevtsova, Lilia. *Lonely Power: Why Russia Has Failed to Become the West and the West Is Weary of Russia*. Washington, DC: Carnegie Endowment for International Peace: 2010.

———. *Putin's Russia*. Washington, DC: Carnegie Endowment for International Peace, 2003.

Shirer, William L. *The Rise and Fall of the Third Reich*. New York: Simon & Schuster, 1960.

Shirk, Susan L. *China: Fragile Superpower*. New York: Oxford University Press, 2007.

Shlapak, David A. *Question of Balance: The Shifting Cross-Strait Balance and Implications for the U.S*. Santa Monica, CA: RAND Corp., 2010.

Siegel, Adam B. "Base Access Constraints and Crisis Response." *Airpower Journal* 10, no. 2 (Spring 1996): 107–12.

Singh, Michael, and Jacqueline Newmeyer Deal. "China's Iranian Gambit." *Foreign Policy*, 31 October 2011. http://www.foreignpolicy.com/articles/2011/10/31/china_iran_nuclear_relationship.

Sokov, Nikolai. "A Second Sighting of Russian Tactical Nukes in Kaliningrad. *CNS*, James Martin Center for Nonproliferation Studies, 15 February 2011. http://cns.miis.edu/stories/110215_kaliningrad_tnw.htm.

Spector, Ronald H. *Eagle against the Sun: The American War with Japan*. New York: Free Press, 1985.

Stockholm International Peace Research Institute. *SIPRI Yearbook 2012*. New York: Oxford University Press, 2012.

Strauss, Barry. *The Battle of Salamis*. New York: Simon & Schuster, 2004.

Summers, Harry G. *Persian Gulf War Almanac*. New York: Facts on File, 1995.

Sun Tzu. *The Art of War*. Translated by Samuel B. Griffith. New York: Oxford University Press, 1963.

Sunday Times of London Insight Team. *War in the Falklands: The Full Story*. New York: Harper & Row, 1982.

Taheri, Amir. *The Persian Night: Iran under the Khomeinist Revolution*. New York: Encounter Books, 2009.

Tan, Andrew T. H., ed. *The Politics of Maritime Power: A Survey*. London: Routledge, 2007.

Tangredi, Sam J. *All Possible Wars? Toward a Consensus View of the Future Security Environment, 2001–2025*, McNair Paper 63. Washington, DC: National Defense University Press, 2000.

———. "The Fall and Rise of Naval Forward Presence." U.S. Naval Institute *Proceedings* 126, no. 5 (May 2000): 28–32.

———. *Futures of War: Toward a Consensus View of the Future Security Environment, 2010–2035*. Newport, RI: Alidade, 2008.

———, ed. *Globalization and Maritime Power*. Washington, DC: National Defense University Press, 2002.

———. "No Game Changer for China." U.S. Naval Institute *Proceedings* 136, no. 2 (February 2010): 24–29.

———. "Sea Basing: Concept, Issues and Recommendations." *Naval War College Review* 64, no. 4 (Autumn 2011): 28–41.

Tangredi, Sam J., and Randall G. Bowdish. "Core of Naval Operations: Strategic and Operational Concepts of the United States Navy." *The Submarine Review*, January 1999: 11–23.

Taylor, Phil, and Pam Cupper. *Gallipoli: A Battlefield Guide*. Kenthurst, Australia: Kangaroo, 1989.

"Text of Newly-Approved Russian Military Doctrine: Report by Russian Presidential Website on 5 February." Military Education Research Library Network, 5 February 2010. http://merln.ndu.edu/whitepapers/Russia2010_English.pdf.

Thornton, Rod. *Asymmetric Warfare: Threat and Response in the Twenty-First Century*. London: Polity, 2007.

Till, Geoffrey. *Seapower: A Guide for the Twenty-First Century*. London: Frank Cass, 2004.

Tiron, Roxana. "China Building Capability to Counter U.S. Defense of Taiwan, Panetta Says." *Bloomberg News*, 7 June 2011. www.bgov.com.

———. "U.S. Places $42 Billion Bet on Carriers in China's Sights." *Bloomberg News*, 19 June 2012. www.bgov.com.

Train, Harry D., III. "An Analysis of the Falkland/Malvinas Islands Campaign." *Naval War College Review* 41, no. 2 (Winter 1988): 33–50.

Turner, Stansfield. "Missions of the U.S. Navy." *Naval War College Review* 26, no.4 (March–April 1974): 2–17.

Uhlig, Frank, Jr. *How Navies Fight: The U.S. Navy and Its Allies*. Annapolis, MD: Naval Institute Press, 1994.

U.S. Commission on National Security/21st Century, *New World Coming*. Washington, DC: U.S. Commission on National Security/21st Century, 1999.

U.S. Congress, Joint Committee on the Investigation of the Pearl Harbor Attack. *Pearl Harbor Attack*, part 12. 79th Cong., 1st sess. Washington, DC: Government Printing Office, 1946.

U.S. Department of Defense. *Annual Report on Military Power of Iran*. Washington, DC: Office of the Secretary of Defense, 2012.

———. *Annual Report to Congress on Military and Security Developments Involving the People's Republic of China*. Washington, DC: Office of the Secretary of Defense. 2010.

———. *Barriers, Obstacles, and Mine Warfare for Joint Operations*. Joint Publication 3–15. Washington, DC: Joint Chiefs of Staff, 2007. http://www.dtic.mil/doctrine/new_pubs/jp3_15.pdf.

———. *Capstone Concept for Joint Operations, Version 3.0*. Washington, DC: Chairman, Joint Chiefs of Staff, 2009. http://www.dtic.mil/futurejointwarfare/concepts/approved_ccjov3.pdf.

———. *Conduct of the Persian Gulf War: Final Report to Congress.* Washington, DC: Office of the Secretary of Defense, 1992. http://www.ndu.edu/library/epubs/cpgw.pdf.

———. *Countering Air and Missile Threats.* Joint Publication 3–01. Washington, DC: Joint Chiefs of Staff, 2007. http://www.dtic.mil/doctrine/new_pubs/jp3_01.pdf.

———. *Department of Defense Dictionary of Military and Associated Terms.* Joint Publication 1–02. Washington, DC: Joint Chiefs of Staff, 2010 (amended through 15 August 2011).

———. *Deterrence Operations Joint Operating Concept (Version 2.0).* Washington, DC: Joint Chiefs of Staff, 2006. http://www.dtic.mil/futurejointwarfare/concepts/do_joc_v20.doc.

———. *Global Strike Joint Integrating Concept (Version 1.0).* Washington, DC: Joint Chiefs of Staff, 2005. http://www.dtic.mil/futurejointwarfare/concepts/gs_jic_v1.doc .

———. *Joint Forcible Entry Operations.* Joint Publication 3–18. Washington, DC: Joint Chiefs of Staff, 2008. http://www.dtic.mil/doctrine/new_pubs/jp3_18.pdf .

———. *Joint Logistics (Distribution) Joint Integrating Concept (Version 1.0).* Washington, DC: Joint Chiefs of Staff, 2006. http://www.dtic.mil/futurejointwarfare/concepts/jld_jic.pdf.

———. *The Joint Operating Environment 2010.* Suffolk, VA: U.S. Joint Forces Command, 2010. http://www.jfcom.mil/newslink/storyarchive/2010/JOE_2010_o.pdf .

———. *Joint Operational Access Concept,* Version 1.0. Washington, DC: Joint Chiefs of Staff, 2012.

———. *Joint Operational Planning.* Joint Publication 5.0. Washington, DC: Joint Chiefs of Staff, 2011.

———. *Joint Operations.* Joint Publication 3–0 (incorporating change 2). Washington, DC: Joint Chiefs of Staff, 2010. http://www.dtic.mil/doctrine/new_pubs/jp3_0.pdf.

———. *Military Contribution to Cooperative Security (CS) Joint Operating Concept (Version 1.0).* Washington, DC: Joint Chiefs of Staff, 2008. http://www.dtic.mil/futurejointwarfare/concepts/cs_jocv1.pdf .

———. *National Defense Strategy.* Washington, DC: Office of the Secretary of Defense, 2008. http://www.defense.gov/news/2008%20national%20defense%20strategy.pdf .

———. *Quadrennial Defense Review.* Washington DC: Office of the Secretary of Defense 1997.

———. *Sustaining U.S. Global Leadership: Priorities for 21st Century Defense.* Washington, DC: Office of the Secretary of Defense, 2012.

———. *A Topic Exploration Paper for Joint Experimentation: Assured Access.* Suffolk, VA: U.S. Joint Forces Command J-9, 2001.

U.S. Naval Institute. "The Maritime Strategy" (supplement). U.S. Naval Institute *Proceedings*, January 1986.

U.S. Navy, Office of Naval Intelligence. *Challenges to Naval Expeditionary Warfare*. Washington, DC: Office of Naval Intelligence, 1997.

UPI.com. "Riyadh Mulls Big Russian Missile Buy." 22 March 2010. www.upi/Business_News/Security-Industry/2010/03/22/Riyadh-mulls-big-Russian-missile-buy/UPI-9311269283852.

Vajdic, Daniel. "Russia's 'Shrewd' Central Asia Play." *The Diplomat*, 17 July 2012. http://thediplomat.com/flashpoints-blog/2012/07/17/russias-shrewd-central-asia-play.

Van Der Vat, Dan. *The Dardanelles Disaster: Winston Churchill's Greatest Failure*. New York: Overlook, 2009.

Van Tol, Jan, with Mark Gunzinger, Andrew Krepinevich, and Jim Thomas. *AirSea Battle: A Point of Departure Operational Concept*. Washington, DC: Center for Strategic and Budgetary Assessments, 2010.

Villahermosa, Gilberto. *Desert Storm: The Soviet View*. Ft. Leavenworth, KS: Foreign Military Studies Office, U.S. Army Command and Staff College, 1992.

Voice of America. "China Calls for "Moderate" Response to North Korean Rocket Launch." 13 December 2012. http://www.globalsecurity.org/wmd/library/news/dprk/2012/dprk-121213-voa01.htm.

Voice of Russia. "China Interested in Russian Missile System." 10 May 2012. http://spacewar.com/reports/China_interested_in_Russian_missile_system-999.html.

Walt, Stephen M. *Taming American Power: The Global Response to U.S. Primacy*. New York: Norton, 2005.

Walton, Timothy A. *China's Three Warfares*. Washington, DC: Delex Systems, 2012.

Waltz, Kenneth. "Why Iran Should Get the Bomb." *Foreign Affairs* 91, no. 4 (July/August 2012), 2–5.

Warden, John A., III. *The Air Campaign: Planning for Combat*. Washington, DC: National Defense University Press, 1988.

———. "The Enemy as a System," *Airpower Journal* 9, no. 2 (Spring 1995): 40–55.

White House, The. *National Security Strategy*. Washington, DC: 2010. http://www.whitehouse.gov/sites/default/files/rss_viewer/national_security_strategy.pdf .

Whitehurst, Clinton H., Jr. *American Military Options in a Taiwan Strait Conflict*. Clemson, SC: Clemson University, 1999.

Williams, Jay. *The Spanish Armada*. New York: American Heritage, 1966.

Winik, Jay. "Secret Weapon: The Pentagon's Andy Marshall Is the Most Influential Man You've Never Heard Of." *Washingtonian*, April 1999: 45–55.

Winterbotham, F. W. *The Ultra Secret*. New York: Harper & Row, 1974.

Wolf, Amy F. *Conventional Prompt Global Strike and Long-Range Ballistic Missiles: Background and Issues*. Report for Congress R41464. Washington, DC: Congressional Research Service, 2012.

Woodward, John, with Patrick Robinson. *One Hundred Days: The Memoirs of the Falklands Battle Group Commander*. Annapolis, MD: Naval Institute Press, 1992.

Worden, Robert L., ed. *North Korea: A Country Study*, 5th ed. Washington, DC: Federal Research Division, Library of Congress, 2008.

Work, Robert O. *The Challenge of Maritime Transformation: Is Bigger Better?* Washington, DC: Center for Strategic and Budgetary Assessments, 2002.

WRITENET. *The Dynamics and Challenges of Ethnic Cleansing: The Georgia-Abkhazia Case, 1 August 1997*. Refworld, United Nations High Commissioner for Refugees. http://www.unhcr.org/refworld/docid/3ae6a6c54.html.

Wylie, J. C. *Strategy: A Theory of Power Control*. New Brunswick, NJ: Rutgers University Press, 1967. Republished with introduction and new postscript. Annapolis, MD: Naval Institute Press, 1989.

Yang Fan. "Terrorism and the 'Unlimited War.'" *Association for Asian Research*, 25 September 2002. http://www.asianresearch.org/articles/915.html.

Yaphe, Judith S., and Charles D. Lutes. *Reassessing the Implications of a Nuclear-Armed Iran*, McNair Paper 69. Washington, DC: National Defense University Press, 2005.

Yoshihara, Toshi, and James R. Holmes. *Red Star over the Pacific: China's Rise and the Challenge to U.S. Maritime Strategy*. Annapolis, MD: Naval Institute Press, 2010.

Index

About the Author

Sam J. Tangredi is a defense strategist whose studies of future warfare prompted Defense Department officials to label him "the Navy's futurist." His thirty-year naval career included command at sea, service in key strategic planning positions in the Pentagon, and research fellowships at two think tanks. He holds a PhD in international relations. His more than one hundred publications have won awards, including the U.S. Naval Institute's Arleigh Burke Prize and the U.S. Navy League's Alfred Thayer Mahan Award. He is currently the director of San Diego operations for the planning/consulting firm Strategic Insight.

The Naval Institute Press is the book-publishing arm of the U.S. Naval Institute, a private, nonprofit, membership society for sea service professionals and others who share an interest in naval and maritime affairs. Established in 1873 at the U.S. Naval Academy in Annapolis, Maryland, where its offices remain today, the Naval Institute has members worldwide.

Members of the Naval Institute support the education programs of the society and receive the influential monthly magazine *Proceedings* or the colorful bimonthly magazine *Naval History* and discounts on fine nautical prints and on ship and aircraft photos. They also have access to the transcripts of the Institute's Oral History Program and get discounted admission to any of the Institute-sponsored seminars offered around the country.

The Naval Institute's book-publishing program, begun in 1898 with basic guides to naval practices, has broadened its scope to include books of more general interest. Now the Naval Institute Press publishes about seventy titles each year, ranging from how-to books on boating and navigation to battle histories, biographies, ship and aircraft guides, and novels. Institute members receive significant discounts on the Press's more than eight hundred books in print.

Full-time students are eligible for special half-price membership rates. Life memberships are also available.

For a free catalog describing Naval Institute Press books currently available, and for further information about joining the U.S. Naval Institute, please write to:

Member Services
U.S. Naval Institute
291 Wood Road
Annapolis, MD 21402-5034
Telephone: (800) 233-8764
Fax: (410) 571-1703
Web address: www.usni.org